T0336978

Advances in Artificial Transportation Systems and Simulation

Advances in Artificial Transportation Systems and Simulation

Edited by

Rosaldo J.F. Rossetti
Ronghui Liu

ELSEVIER

AMSTERDAM • BOSTON • HEIDELBERG • LONDON
NEW YORK • OXFORD • PARIS • SAN DIEGO
SAN FRANCISCO • SINGAPORE • SYDNEY • TOKYO

Academic Press is an imprint of Elsevier

Academic Press is an imprint of Elsevier
525 B Street, Suite 1800, San Diego, CA 92101-4495, USA
225 Wyman Street, Waltham, MA 02451, USA
The Boulevard, Langford Lane, Kidlington, Oxford OX5 1GB, UK

Copyright © 2015 Zhejiang University Press Co., Ltd. Published by Elsevier Inc. All rights reserved.

No part of this publication may be reproduced or transmitted in any form or by any means, electronic or mechanical, including photocopying, recording, or any information storage and retrieval system, without permission in writing from the publisher. Details on how to seek permission, further information about the Publisher's permissions policies and our arrangements with organizations such as the Copyright Clearance Center and the Copyright Licensing Agency, can be found at our website: www.elsevier.com/permissions.

This book and the individual contributions contained in it are protected under copyright by the Publisher (other than as may be noted herein).

Notices
Knowledge and best practice in this field are constantly changing. As new research and experience broaden our understanding, changes in research methods, professional practices, or medical treatment may become necessary.

Practitioners and researchers must always rely on their own experience and knowledge in evaluating and using any information, methods, compounds, or experiments described herein. In using such information or methods they should be mindful of their own safety and the safety of others, including parties for whom they have a professional responsibility.

To the fullest extent of the law, neither the Publisher nor the authors, contributors, or editors, assume any liability for any injury and/or damage to persons or property as a matter of products liability, negligence or otherwise, or from any use or operation of any methods, products, instructions, or ideas contained in the material herein.

British Library Cataloguing-in-Publication Data
A catalogue record for this book is available from the British Library

Library of Congress Cataloging-in-Publication Data
A catalog record for this book is available from the Library of Congress

ISBN: 978-0-12-397041-1

For information on all Academic Press publications
visit our website at http://store.elsevier.com/

Typeset by Thomson Digital

Printed and bound in the United States

Working together
to grow libraries in
developing countries

www.elsevier.com • www.bookaid.org

Table of Contents

List of Contributors

Fábio Aguiar   MIEIC, DEI, Faculdade de Engenharia da Universidade do Porto, Porto, Portugal

João Emílio Almeida   LIACC, DEI, Faculdade de Engenharia da Universidade do Porto, Porto, Portugal

Jean-Paul A. Barthès   Université de Technologie de Compiègne, UMR CNRS Heudiasyc, France

Ana L.C. Bazzan   Instituto de Informatica, UFRGS, Porto Alegre, RS, Brazil

Philippe Bonnifait   Université de Technologie de Compiègne, UMR CNRS Heudiasyc, France

Sara Carvalho   MIEIC, SAPO Labs, Faculdade de Engenharia da Universidade do Porto, Porto, Portugal

António J.M. Castro   LIACC, Faculdade de Engenharia da Universidade do Porto, Porto, Portugal

Paul Davidsson   Department of Computer Science, Malmö University, Malmö, Sweden

Manoel T. de Abreu Netto   Department of Computer Science, PUC-Rio, Rio de Janeiro, RJ, Brazil

Maicon de Brito do Amarante   Instituto Federal Farroupilha, São Vicente do Sul, RS, Brazil

Carlos J.P. de Lucena   Department of Computer Science, PUC-Rio, Rio de Janeiro, RJ, Brazil

Baldoino F. dos Santos Neto   Department of Computer Science, PUC-Rio, Rio de Janeiro, RJ, Brazil

João Filguieras   Instituto de Engenharia de Sistemas e Computadores, INESC-ID, Lisbon, Portugal

Joaquim Gabriel   IDMEC, DEMec, Faculdade de Engenharia da Universidade do Porto, Porto, Portugal

John Graham   Complex Systems and Non-linear Dynamics Group, Universidad Autónoma de la Ciudad de México, San Lorenzo, Del Valle, México, D.F. México

Shaza Hanif   Department of Computer Science, KU Leuven

Milton Heinen   Instituto de Informatica, UFRGS, Porto Alegre, RS, Brazil

Sergio Hernandez   Postgraduation Program in Complex Systems and Non-linear dynamics, Universidad Autónoma de la Ciudad de México, San Lorenzo, Del Valle, México, D.F. México

Johan Holmgren   Faculty of Computing, Blekinge Institute of Technology, Karlshamn, Sweden, Department of Computer Science, Malmö University, Malmö, Sweden

Tom Holvoet   Department of Computer Science, KU Leuven

Zafeiris Kokkinogenis   LIACC, DEI, Faculdade de Engenharia da Universidade do Porto, Porto, Portugal

Zhengjiang Li   State Key Laboratory of Management and Control for Complex Systems, Institute of Automation, Chinese Academy of Sciences, Beijing, China

Nuno Machado   MIEIC, DEI, Faculdade de Engenharia da Universidade do Porto, Porto, Portugal

Peter T. Martin   Department of Civil Engineering, New Mexico State University, Las Cruces, New Mexico, USA

Antonio Neme   Complex Systems and Non-linear Dynamics Group, Universidad Autónoma de la Ciudad de México, San Lorenzo, Del Valle, México, D.F. México

Omar Neme   School of Economics, Instituto Politécnico Nacional, México, D.F. México

Eugénio Oliveira   LIACC, DEI, Faculdade de Engenharia da Universidade do Porto, Porto, Portugal

Lúcio Sanchez Passos   LIACC, DEI, Faculdade de Engenharia da Universidade do Porto, Porto, Portugal

Linda Ramstedt   Sweco, Stockholm, Sweden

Rosaldo J.F. Rossetti   LIACC, DEI, Faculdade de Engenharia da Universidade do Porto, Porto, Portugal

Luís Sarmento   LIACC, SAPO Labs, Faculdade de Engenharia da Universidade do Porto, Porto, Portugal

Ivana Tasic   Department of Civil and Environmental Engineering, University of Utah, Salt Lake City, Utah, USA

Fenghua Zhu   State Key Laboratory of Management and Control for Complex Systems, Institute of Automation, Chinese Academy of Sciences, Beijing, China

Milan Zlatkovic   Department of Civil and Environmental Engineering, University of Utah, Salt Lake City, Utah, USA

Preface

Intelligent Transportation Systems (ITS) have evolved enormously within the last three decades. More recently the rapid growth in new technologies has allowed for the practical implementation of more interactive and pervasive ITS solutions. Because of the important role the transportation systems play in society and economy, industry has engaged actively as a great promoter of ITS' technological development and governments all over the world are prioritizing mobility as a key ingredient of their social and economic growth. Increasingly, ITS developments become not solely technological, but are across a wide span of different disciplines.

Despite of the exponential growth in computer power and communication technologies underlying ITS, meeting the future challenges in transportation will require focusing on social and environmental aspects where user preferences are a central concern. Human interactions with ITS solutions have now gained a new meaning. As technology is able to behave more intelligently than before, services become peers of users: they perceive, make decisions and reason about the results of their actions, all the while seeking to benefit all parties. In fact, rather than increasing service capacity, one underlying approach of ITS-based solutions nowadays is to ensure productivity and mobility by making better use of existing infrastructure and services, furnishing them with smarter, greener, safer, and more efficient solutions.

The complexity of contemporary ITS technology and its wider social and environmental impacts demand new modeling paradigms that incorporate cooperation and collaboration among intervening parties, and support the design and practical deployment of future mobility solutions. In response to the need to understand the interplay between technology and social interactions, Prof. Fei-Yue Wang proposed the concept of Artificial Transportation Systems (ATS), just over a decade ago, during the 2003 IEEE International Conference on Intelligent Transportation Systems. With the ability to integrate different transportation models and solutions in a virtual environment, ATS are an extension to traditional modeling and simulation methodologies that deal with transportation issues from the complex systems perspective and in a systematic and synthetic way. They provide a natural platform where new approaches can be experimented while avoiding natural drawbacks of dealing directly with real-life critical domains. Building on the theories and metaphors developed in a wide spectrum of disciplines, spanning from social sciences, artificial intelligence, and multi-agent systems, to distributed computing and virtual reality, many important issues arise in ATS that challenge and motivate researchers and practitioners from multidisciplinary technical and scientific backgrounds.

Inspired by the concept, IEEE ITS Society soon after created and has since hosted the Technical Activities Sub-committee on Artificial Transportation Systems and Simulation (ATSS) with the mission of motivating and promoting research and practice in ATSS. As part of this effort, a series of successful biennial ATSS Workshops has been organized by this committee and integrated in scientific programs of IEEE ITSC, the flagship conference series of the society. In addition, every other year, alternating with ATSS workshops, a series of special sessions have been organized as part of ITSC to consolidate ideas and trends discussed in previous ATSS workshops. This book is a major outcome of the ATSS sub-committee, bringing to the readers a collection of selected papers presented during the ATSS Workshop held in ITSC 2010, in Madeira, Portugal, and during the ATSS Special Session held in ITSC 2011, in Washington-DC, USA. The papers herein included report on important aspects and technological advances underlying the concept of ATS and establishing the basis for the implementation of appropriate simulation methodologies and tools.

Each chapter in this book consists of an extended and updated version of papers previously presented in the above two ATSS events. A total of 12 papers have been selected that cover different aspects of ATS and span across topics such as simulation tools, modeling methodologies, practical applications, alternative data sources, crowd sensing, and participatory simulation. Starting with tools, Bazzan et al. present ITSUMO, an open-source microscopic traffic simulator whose implementation relies on the agent metaphor. Differently from other similar tools, ITSUMO integrates both demand and control perspectives. It follows a bottom-up behavioral approach, offering sufficient flexibility for the development of different algorithms and techniques to test with route assignment and re-planning, driver behavior, and traffic control strategies and coordination.

Netto et al. discuss issues related to the representation of complex domains, and propose a framework as a reusable solution for building decentralized self-organizing systems, based on major architectural patterns found in the literature. Considering a higher level of abstraction, their approach provides extensibility features to develop new interaction and coordination mechanisms between agents and the environment, which is demonstrated in an automated guided vehicles scenario. Passos et al. also tackle complex system analysis from a multi-agent perspective. In their work, authors analyze the adequacy of traditional approaches in the field of Agent-Oriented Software Engineering to create adequate multi-agent systems in the specific domain of transportation. They devise a novel methodology where the concept of services is considered as peer of agents, ambience, and processes, and becomes prominent elements in the modeling phase. The approach is illustrated in a typical transportation domain scenario. Holmgren et al. explore modeling specificities in the supply chain domain. The authors propose a method based on a framework of supply chain roles, responsibilities, and interactions, which can be used to represent different types of organizations involved in providing and using products and transport services. Their method is illustrated through five different supply chain simulation models, which are analyzed to demonstrate validity and generality of their approach. Also talking about modeling issues, Hanif and Holvoet illustrate how design patterns can support the design of complex agent-based solutions to Pickup and Delivery Problems. In particular, authors use the so-called delegate multi-agent-system patterns to build agent interaction behaviors, and show through simulation that a structured and reusable pattern can significantly reduce the system design and implementation complexity, yet achieve interesting quality characteristics.

Reporting on the applications, Machado et al. demonstrate how the ATS concept can be used in practice as means to assess and evolve the organizational structure performance of airline companies. The authors build on the empirical knowledge gained through interviews with airline operators and develop an analytical framework so as to evaluate current as well as hypothetical organizational structures. To illustrate their approach, real pre- and postoperational data is used to support the simulation of different operation scenarios allowing for different metrics to be analyzed. Neme et al. address issues in modeling pedestrian dynamics and study passengers inside high-capacity buses. Through ATS they observe uneven density distributions leading to high discomfort to the passengers. An agent-based model was devised to represent the interactions between passengers and the bus interior, comprising seats, aisle, and access doors. The authors analyze different schemas and policies so as to gain insight into how to improve passenger comfort on buses. In a different perspective, Almeida et al. study pedestrian dynamics in relation to the behavior of crowds in emergency situations, and present ModP, an agent-based pedestrian simulator. The tool is flexible enough to allow different behaviors to be modeled through a simple syntax, providing designers with a productive environment to rapidly prototype and test with different scenarios. The authors carried out preliminary studies to illustrate usability of their tool.

The implementation of cognitive capabilities in intelligent vehicles interfaces is addressed by Barthès and Bonnifait as a means to allow vehicles to interact collaboratively with their drivers in operating conditions. The interactive interface is built on top of a multi-agent system and tested in an Advanced Driving Assistance System scenario providing speed warnings whenever dangerous areas are approached. The authors test their approach in a real-world scenario resorting to an experimental vehicle. Zhu and Li look into new methods to address transportation problems from new perspectives using the Artificial Societies, Computational Experiments and Parallel Execution approach, built upon the ATS premises. The authors emphasize on synthesizing artificial societies and modeling environmental impacts, whereas their architectural approach relies on a cloud computing infrastructure. They illustrate the concept with an ATS case study and present preliminary results of their methodology. In addition, Zlatkovic et al. illustrate one important premise of ATS, namely its ability to allow for software- and hardware-in-the-loop simulations. Their work present and discuss an implementation of software-in-the-loop (SIL) simulation of the Advanced System Controller series 3 (ASC/3) in transit signal priority scenarios. The authors test two options of ASC/3 using a VISSIM simulation model of a bus rapid transit solution in West Valley City, Utah. Results are encouraging and demonstrate how SIL simulation can offer many options for testing custom-defined traffic control strategies.

Alternative data sources, crowd-sensing, and participatory simulation are also topics of major concern in ATS. Kokkinogenis et al. discuss a new type of mobility studies resorting to the growth in popularity of opinion mining in social media, and social media itself. Users are considered to be sensors of the mobility dynamics, capable of providing insight into the flaws of mobility networks, user preferences, and other unexplored sorts of information. Their approach builds upon sensing real-time traffic-related information using microblogging messages posted on Twitter (by users in transit), to which a text classification approach is proposed. This work opens up vast swathes of opportunities to explore alternative data sources leveraged on mass participation and collaboration.

We are very pleased with this interesting and motivating collection of papers reporting on the basic concepts, state-of-the-art developments, and a myriad of potential applications and future trends in Artificial Transportation Systems and Simulation. We hope this book will serve as a source of inspiration and insight for many researchers, practitioners, and educators, and leverage further progresses in the field of ATSS. We conclude with a word of appreciation to all authors who have submitted their contributions to ATSS events (in both workshops and special sessions) throughout the past 10 years, with clear visions, novel research, and significant results. Finally, the contributions included in this book are results of laborious and time-consuming work of many reviewers who have helped us with their expertise, suggestions, and recommendations, and to whom we are greatly indebted. Thank you!

October 2014

Rosaldo J.F. Rossetti

Laboratório de Inteligência Artificial e Ciência de Computadores,
Departamento de Engenharia Informática,
Faculdade de Engenharia da Universidade do Porto,
Rua Dr. Roberto Frias, S/N, Porto, Portugal

Ronghui Liu

Institute for Transport Studies, University of Leeds, 34–40 University Road,
Leeds, United Kingdom

CHAPTER 1

ITSUMO: An Agent-Based Simulator for Intelligent Transportation Systems

Ana L.C. Bazzan*, Milton Heinen**, Maicon de Brito do Amarante†

*Instituto de Informatica, UFRGS, Porto Alegre, RS, Brazil; **Instituto de Informatica, UFRGS, Porto Alegre, RS, Brazil; †Instituto Federal Farroupilha, São Vicente do Sul, RS, Brazil

1.1 Introduction and Motivation

The second half of the last century has seen the beginning of the phenomenon of traffic congestion. This arose due to the fact that the demand for mobility in our society has increased constantly. Traffic congestion is a phenomenon caused by too many vehicles trying to use the same infrastructure at the same time. The consequences are well known: delays, air pollution, decrease in speed, and dissatisfaction (which may lead to risk maneuver thus reducing safety for pedestrians as well as for other drivers).

The increase in transportation demand can be met by providing additional capacity. However, this might no longer be economically or socially attainable or feasible. Thus, the emphasis has shifted to improving the existing infrastructure without increasing the overall nominal capacity, by means of a better utilization of this capacity. Two complementary measures can be taken. In traffic engineering terminology these are associated with management of the demand (users, drivers) and supply (infrastructure, control). The set of all these measures is framed as Intelligent Transportation Systems (ITS).

In the last years there have been some proposals for simulation platforms that are flexible enough to test ITS techniques and approaches. Some (e.g., Paramics, AISUM, VISIM, EMME2, Dracula) are based on classical models of simulation and are commercial tools. With the appearance of a new simulation paradigm – agent-based simulation – it is now possible that traffic experts and other users develop their own applications. This has been achieved to some extent (e.g., Dresner and Stone, 2004; Rossetti and Liu, 2005; van Katwijk et al., 2005; Balmer et al., 2008; Bazzan et al., 1999; Burmeister et al., 1997; Tumer et al., 2008; Vasirani and Ossowski, 2009, 2011) but these tools are goal-directed meaning that they were built for (more or less) specific purposes. One of the notable exceptions is MATSim (www.matsim.org). However, MATSim's simulation paradigm is queue based, traffic lights are very simple, and drivers are not fully autonomous (e.g., during replanning). We remark that besides the commercial simulators mentioned above, there is also the possibility to use SUMO (http://sumo.sourceforge.net/) as a starting point to investigate traffic scenarios using microscopic traffic simulations. However, SUMO does not yet allow native tools for implementing agent-based solutions.

In short, most of previously mentioned works have one or more of the following drawbacks: they are not fully agent based; they rely on strong simplifying assumptions; they do not consider both control and assignment of demand as a whole process (except in (Vasirani and Ossowski, 2009, 2011), but here the integration only refers to their specific market-based approach).

Therefore, there is a lack of support for traffic experts who want to implement and test their own solutions (e.g., artificial intelligence (AI)-based approaches for optimization or broadcast of recommendation). These experts can neither extend commercial tools (except for some API-based modules, which (1) are not totally flexible and (2) represent an additional purchase cost), nor use the available free tools as they deal only with pieces of the whole problem.

This way there is still the need for an integrated platform that is as follows: fully based on the autonomous agent paradigm of simulation; open source and user friendly; and considers the effects of both control measures on driver's reasoning and vice-versa. The present chapter describes ITSUMO (Intelligent Transportation System for Urban Mobility), an open-source tool that addresses these issues. It allows the modeling of traffic actors (drivers, traffic lights, and even autonomous vehicles) as autonomous agents; it deals with short-term control of traffic lights and with en-route replanning by drivers; thus it permits the study of coeffects of both demand and supply. This is achieved by means of AI techniques in general and of agent-based techniques in particular.

With the increased dissemination and computing power of mobile devices, it is now possible to execute distributed AI applications for various situations: intelligent routing using algorithms that do not rely on full knowledge; planning under constraints, and restricted communication and information; distributed optimization of traffic lights. For instance, it is possible to define drivers as intelligent agents and to plug each driver model. This approach is different from current models, which are purely reactive and ignore drivers' mental states (informational and motivational data). Also, it is possible to plug reinforcement-learning-based control for traffic lights.

An earlier version of ITSUMO was presented as a demo in the AAMAS conference (da Silva et al., 2006b). However, although ITSUMO has also been used to investigate route choice scenarios, the focus has been primarily on control. The current version was extended in the sense that it now allows modeling of both control measures and drivers reaction to them, as well as routing techniques. Moreover, this is provided as noncommercial code and is one of the few tools that are truly agent based (thus microscopic). As shown in the next section, the simulation kernel is responsible for handling the movement of vehicles. Other modules support the agent-based modeling of demand and supply.

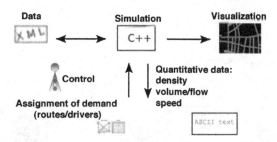

Figure 1.1: High-level view of the simulation modules

In the next section details of the simulator and an overview of the main modules are given. Afterwards, two of these modules are presented in detail: the one that regards control (Section 1.3), and the one that refers to demand (Section 1.4). Section 1.5 revisits the main aspects by means of a case study where traffic lights adapt and where autonomous drivers can plan their routes. The last section concludes the chapter.

1.2 Description of the Simulator

ITSUMO is composed of five modules: database, the simulation kernel, control, demand (assignment and drivers' definition), and the output module (visualization and statistics). Figure 1.1 shows how these modules interact.

In order to run a simulation, the topology must be stored as an XML file. After running the simulation, two optional outputs can be used: either through on-screen visualization (macroscopic or microscopic) or via dump of various data files. Other optional modules are the insertion and control via signal plans, and the assignment.

Figure 1.2 provides more details about specific functionalities that will be discussed in the next sections. As discussed in the following sections, ITSUMO allows data configuration in various ways, and also provides the basic interfaces for its extension. For example, it is possible to extend the framework adding new routing algorithms or new traffic light control methods. The current alternatives are as in Table 1.1.

1.2.1 Microscopic Simulation Model and Simulation Kernel

In contrast to macroscopic models of traffic simulation (which are mainly concerned with the movement of platoons of vehicles, focusing on the aggregate level), in the agent-based paradigm each object can be described as detailed as desired, thus permitting a more realistic modeling of drivers' behavior for instance. In the agent-based approach both for travel and/or route choices may be considered, which is a key issue in simulating traffic since those choices are becoming increasingly more complex. Also, individual traffic lights can be modeled according to several approaches, from classical off-line coordination to recently proposed

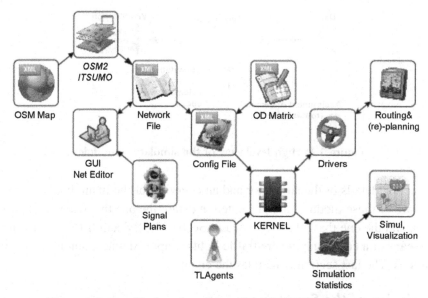

Figure 1.2: Specific functionalities of the ITSUMO framework

ones (negotiation, communication free, via game theory, reinforcement learning, swarm intelligence, etc.)

In order to achieve the necessary simplicity and performance, ITSUMO uses the Nagel–Schreckenberg cellular-automata (CA) model (Nagel and Schreckenberg, 1992) for traffic movement (aka. Na-Sch model). In short, each road is divided into cells with a fixed length.

Table 1.1: Alternatives for topology and traffic light edition, control, routing algorithms, conditions for routing, and planning.

Topology		Manual	
		OSM	
Traffic light generation		Manual	
		Automatic	
Traffic light control		Fixed time	
		Greedy	
		RL	
Routing algorithms		Dijkstra	
		A*	
		ARA*	
Routing, planning, and replanning	Pretrip	OD-based	
		Na-Sch + FC	
		Na-Sch	
	En-route	Na-Sch	
		Congestion-based	Full knowledge
			Partial knowledge

This allows the representation of a road as an array where vehicles occupy discrete positions. The movement follows rules that represent a special form of car-following behavior. This simple, yet valid microscopic traffic model can be implemented in such an efficient way that is good enough for real-time simulation and control of traffic.

Hence the kernel of the simulator (implemented in C++) is based on the CA model. The simulation occurs in discrete steps and is implemented as a series of updates in the vehicles' positions in the network. Each update in a node or traffic light may modify its previous behavior.

1.2.2 Database Module

The information regarding the topology of the traffic network is stored in an XML file (see Figure 1.3). The database module creates, updates, and stores the static and the dynamic objects to be used in the simulation, both related to the infrastructure (supply) and to the demand. Regarding the former, the main attributes are: Cartesian coordinates of intersections, streets characteristics (number of lanes, etc.), and signal plans (set of lane-to-laneset allowed movements). Regarding the demand, the dataset stores the following: insertion rate of vehicles at given nodes of the network; origin and destination of drivers, and so forth. Here we show the attributes related to the topology only (see Figure 1.4). For more details about these attributes please see (da Silva et al., 2006b).

As indicated at the top of Table 1.1, this kind of data can be either entered manually – via a GUI, or be imported directly from the Open Street Map (OSM) (www.openstreetmap. org) through an application known as "OSM2ITSUMO" (see Figure 1.2). Using the former method, each component in the network (e.g., as depicted in Figure 1.4 for a small example) is inserted by the user. This process is time consuming and error prone. It can however be used for small networks. For complex or big maps, the alternative is to use the latter method.

Because the XML format used in OSM is different from the one used in ITSUMO, we provide a parser to get the necessary data from OSM. The user just has to select the portion of the OSM map he/she wants to use (typically a bounding box as in Figure 1.5), export this to an XML (in OSM format), and run the parser. The output is then a new XML in the ITSUMO format (i.e., nodes, streets, sections, lanesets, and lanes are created with their corresponding attributes).

It is important to remark that most of the streets in OSM have a tag to classify them (motorway, primary, secondary, residential, etc.). Thus, our parser can import streets that match one or more of these tags. For instance, it is possible to import just the main arterials present in an OSM map, or all links, or links in any other degree of abstraction. Hence, by importing OSM maps (in different abstraction levels), ITSUMO permits the use of real-world maps.

```
            <nodes>
                <node>
                    <node_id> 2 </node_id>
                    <node_name> n0 </node_name>
                    <x_coord> 6250.0 </x_coord>
                    <y_coord> 6000.0 </y_coord>
                </node>
                ...
        <traffic_lights>
            <traffic_light>
                <traffic_light_id> 161 </traffic_light_id>
                <located_at_node> 14 </located_at_node>
                <signalplans>
                    <signalplan>
                        <signalplan_id> 1 </signalplan_id>
                        <phases>
                        <phase>
                            <phase_id> 1 </phase_id>
                            <iteration_start> 0 </iteration_start>
            ...
        <section>
            <section_id> 74 </section_id>
            <section_name> (g2) &lt; -&gt; (n2) </section_name>
            <is_preferencial> false </is_preferencial>
            <delimiting_node> 25 </delimiting_node>
            <delimiting_node> 4 </delimiting_node>
            <lanesets>
                <laneset>
                    <laneset_id> 75 </laneset_id>
                    <laneset_position> 1 </laneset_position>
                    <start_node> 25 </start_node>
                    <end_node> 4 </end_node>
                    <turning_probabilities>
                        <direction>
                            <destination_laneset> 78 </destination_laneset>
                            <probability> 100.0 </probabili ty>
                        </direction>
                    </turning_probabilities>
```

Figure 1.3: XML file (partially) describing a traffic network in ITSUMO (*Continued*)

```
                    <lanes>
                        <lane>
                            <lane_id> 76 </lane_id>
                            <lane_position> 1 </lane_position>
                            <maximum_speed> 10 </maximum_speed>
                            <deceleration_prob> 0.0
                            </deceleration_prob>
                        </lane>
                    </lanes>
                </laneset>
            </lanesets>
        </section>
        ...
```

Figure 1.3: (*Continued*)

Similar to the definition of the network topology, there are two ways to create signal plans in ITSUMO (notice, however, that the use of traffic lights is optional since the simulation can be run without defining traffic lights and corresponding plans). As indicated in Table 1.1, plans can be created manually or automatically. In the former, the user is requested to enter the full definition of all signal plans for each signalized intersection. Figure 1.6 depicts the GUI provided for creation/edition. For each plan, phases must be defined (using the mouse to connect incoming lanes to outgoing lanesets), their cycle times, and splits (not shown). This is again time consuming. If the network has too many signalized intersections, ITSUMO's automatic signal plan generator can be used. The user has to inform the cycle time, which will be divided equally among all defined movements. This of course generates only one simplified plan. However, users can edit the plan (using the same GUI used for manual edition) and modify it as desired.

The database also stores other objects such as sources, sinks, turning probabilities, and so forth. Due to lack of space we refer reader to elsewhere for more details (da Silva et al., 2006b).

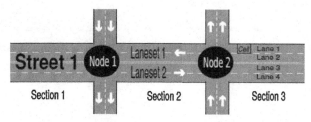

Figure 1.4: ITSUMO: main objects of the topology

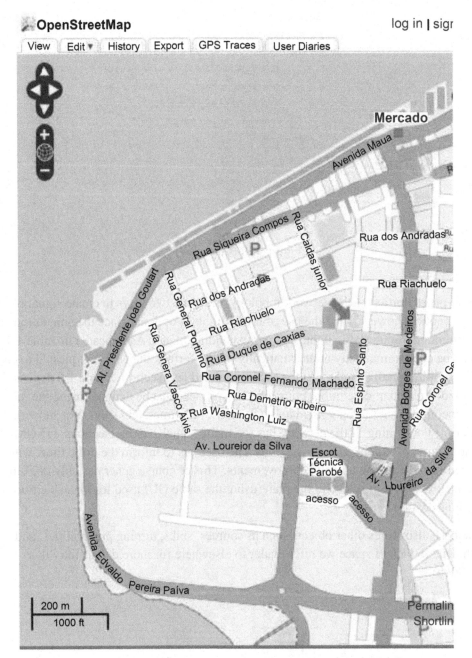

Figure 1.5: OSM export feature (here for Porto Alegre, Brazil)

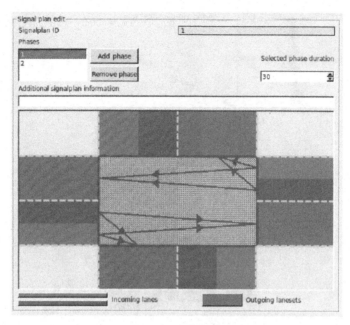

Figure 1.6: GUI for editing signal plans

1.2.3 Output Module: Statistics and Visualization

Sensors and detectors are used to collect information that is displayed during the simulation. Thus, sensors collect all sorts of information about the scenario being simulated, such as the lane occupation rate, the average vehicle speed in a street, in/out flow of vehicles in a specific laneset, and so forth.

The simulation output can be formatted according to the user needs. The most usual formats are the "cell map" and the "laneset occupation map." The former indicates portions of the lane that are occupied by which vehicle, providing the most detailed output possible. On the other hand, the "laneset occupation map" is a high-level output that specifies the rate of occupation (density) for each laneset in the network.

Users can visualize the simulation either at a macroscopic or at a microscopic level (individual vehicles). Both can be seen in Figure 1.7 where the small figure at top left shows the microscopic visualization of one intersection. At a macroscopic level, the visualization considers only data that reflects the overall behavior of the network, providing a useful tool to capture the big picture of what is happening in a specific scenario. The microscopic level provides an interface through which one can see individual vehicles movement. A third kind of visualization is a plot of a subset of the vehicles route across a map using Google Maps.

Figure 1.7: Macroscopic and microscopic visualization of a simulation

1.3 Control: Traffic Light Agent Module

In ITSUMO the control of traffic lights is implemented and executed via traffic light agents. These can control one or more intersections, using the same kind of control method or different ones. This is so because each agent is independent, that is, decides its own action taking into account the information available and using its own algorithm. Notice however that the information available may come from other traffic light agents so that it is not the case that each agent uses exclusively local information. In summary, traffic light agents may be heterogeneous and handle congestion in different ways.

Some basic classes for creating traffic light agents ("TL Agents" in Figure 1.2) have been implemented in order to facilitate the development of traffic controllers. These agents are organized in a data structure that is kept separated from the simulator. Thus, the user does not need to manipulate the kernel code. Moreover, if any user wishes to code its own control method, this can be easily done.

A communication is established between the agents and the kernel using sockets. This permits the exchange of information about traffic status and control actions. The former can be, for example, number of stopped vehicles, density, speed etc. of the lanes under control by the given agent. The agent can then send a control action back to the simulation kernel (normally this control action is the ID of the signal plan that should be run at a given intersection).

So far we provide the following control methods: fixed time (i.e., only one signal plan is used and no change is made so that in fact no control is actually performed); greedy; reinforcement learning based approaches; and an approach based on swarm intelligence.

1.3.1 Greedy Traffic Light Agent

The basic idea underlying the greedy strategy for traffic lights control is to provide more green time to the most congested direction. Currently this is implemented in ITSUMO in two ways: greedy selection of signal plan, or greedy modification of the current plan. In the former, a set of signal plans must be created a priori. The controller then selects the one that provides more green time to the movement that has the longest queue. In the latter, the current plan (which can be the only one defined) is modified so that the movement that has the longest queue gets more green time. The magnitude of the increase in green time is given by a factor (e.g., 10%); the determination of the longest queue is made over a given time window (e.g., the last 5 min).

1.3.2 Reinforcement Learning-Based Methods

Usually, Reinforcement Learning (RL) problems are modeled as Markov Decision Processes (MDPs). These are described by a set of states, S, a set of actions, A, a reward function $R(s, a) \rightarrow \Re$ and a probabilistic state transition function $T(s, a, s') \rightarrow [0, 1]$. An experience tuple $\langle s, a, s', r \rangle$ denotes the fact that the agent was in state s, performed action a and ended up in s' with reward r. Given an MDP, the goal is to calculate the optimal policy π^*, which is a mapping from states to actions such that the discounted future reward is maximized.

In ITSUMO, both model-free and model-based approaches were implemented. *Model-free* systems, such as Q-learning do not require agents to have access to information about how the environment works. Q-Learning works by estimating state-action values, the Q-values, which are numerical estimators of quality for a given pair of state and action. More precisely, a Q-value Q(s, a) represents the maximum discounted sum of future rewards an agent can expect to receive if it starts in state s, chooses action a and then continues to follow an optimal policy. Q-Learning algorithm approximates Q(s, a) as the agent acts in a given environment. If all pairs state-action are visited during the learning process, then Q-learning is guaranteed to converge to the correct Q-values with probability one (Watkins and Dayan, 1992). When the Q-values have nearly converged to their optimal values, the action with the highest Q-value for the current state can be selected.

This approach is implemented in ITSUMO and was used mostly as basis of comparison with other methods, as, for example, in (Bazzan et al., 2010b; de Oliveira and Bazzan, 2009; de Oliveira et al., 2006). In particular, ITSUMO also supports various levels of control as in (Bazzan et al., 2010b) where there is a layered architecture where some agents supervise a group of other agents.

In *model-based* approaches, traffic flow patterns are dynamic and nonstationary in nature. Thus one solution is to keep multiple models of the environment (and their respective policies). These models can be incrementally built. One approach for this appears in (da Silva et al., 2006a) and is denominated RL-CD for Reinforcement Learning with Context Detection. In RL-CD it is assumed that: (1) environmental changes are restricted to a small number of contexts (traffic patterns), which are stationary environments with distinct dynamics; (2) the current context cannot be directly observed, but can be estimated according to the types of transitions and rewards observed; (3) the environmental context changes are independent of agent's actions; and (4) context changes are relatively infrequent. This method automatically partitions the environment dynamics into relevant partial models. Each model is assigned to an optimal policy (which is a mapping from traffic patterns to signal plans), and to a trace of prediction error of transitions and rewards, aiming at estimating the quality of a given partial model. The creation of new models is controlled by a continuous evaluation of the prediction errors generated by each partial model. A partial model contains estimated transition and estimated reward functions.

Incremental Gaussian Mixture Network (IGMN) was discussed in (Heinen, 2011; Heinen and Engel, 2010) as a new approximation function that can be used to approximate continuous states and actions in RL applications. The network is composed of an association region and several cortical regions. All regions have the same number of neurons. Initially there is a single neuron in each region, but more neurons are incrementally added when necessary based on an error driven mechanism. Each cortical region N^K receives signals from the kth sensory/motor modality, k.

IGMN has two operation modes, called *learning* and *recalling*. However, unlike most ANN models, in IGMN these operations do not need to occur separately, that is, the learning and recalling modes can be interwoven. In fact, even after the presentation of a single training pattern the neural network can already be used in the recalling mode (the acquired knowledge can be immediately used), and the estimates become more precise as more training data are presented. Moreover, the learning process can proceed perpetually, that is, the neural network parameters can always be updated as new training data arrive.

This type of RL algorithm is already implemented in ITSUMO, that is, the user only needs to define and use three cortical regions, N^S, N^A and N^Q, to represent the states, s, actions, a, and the Q(s, a) values, respectively.

In order to illustrate its use, we employ this algorithm in the experiments reported in Section 1.5, where we show a case study. A new strategy for selecting continuous actions is employed. This strategy consists in first propagating through the IGMN network the current state, s, and the maximum value, Q_{max}, currently stored in the corresponding Gaussian units. This is formalized as in equation (1.1) where M is the number of neurons in the neural network.

$$Q_{max} = \max_{j \in M} (\mu_j^Q) \qquad (1.1)$$

The Q_{max} value is then propagated through the cortical region N^O, the associative region P is activated and the *greedy* action $\hat{a}$ is computed in the cortical region N^A.

For action selection, instead of simply choosing the greedy action $\hat{a}$ at each moment we can select the actions randomly using the estimated covariance matrix C^A, that is, the actions can be selected randomly using a Gaussian distribution of mean $\hat{a}$ and covariance matrix C^A. In the beginning of the learning process, when M = 0, the initial action can be randomly chosen.

1.3.3 Swarm-Intelligence Inspired Signal Plan Choice

In Bonabeau et al. (Bonabeau et al., 1999) a mathematical model is presented, which formalizes a hypothesis of how the division of labor may happen in colonies of social insects. Interactions among members of the colony and the individual perception of local need result in a dynamic distribution of tasks. Their model describes the colony task distribution using the stimulus produced by tasks that need to be performed and an individual response threshold related to each task. Each individual insect has a response threshold for each task to be performed. That means, at individual level, each task has an associated stimulus, like for instance the perception of waste as a stimulus for cleaning behavior.

The levels of the stimulus increase if tasks are not performed, or not performed by enough individuals. An individual that perceives (e.g., after walking around randomly) a task stimulus higher than its associated threshold has a higher probability to do this task. This model also includes a simple way of reinforcement learning where individual thresholds decrease when performing some task and increase when not performing. This double-reinforcement process leads to the emergence of specialized individuals.

These concepts are implemented in ITSUMO (encapsulated as a traffic light agent) in the following way: each agent (traffic light/crossing) has a social-insect behavior. It has different tendencies to execute one of its signal plans (each signal plan is considered an available task), according to the environment stimulus and particular thresholds. Beside these individuals, this approach also considers that each vehicle leaves a pheromone trace that can be perceived by the agents at the junction. This metaphor is realistic since many junctions have loop induction sensors which detect the counting of vehicles (and sometimes speed).

Signal plans are seen as tasks to be performed by the insect without any centralized control or task allocation mechanism. Stimuli to perform or to change tasks are provided by vehicles that, while waiting for their next green indication, continuously produce "pheromone." Thus the volume of traffic coming from one direction can be evaluated by the intersection agent and this may trigger some signal plan switching. No other information is available to agents. For more details and results, the reader is referred to (de Oliveira and Bazzan, 2006).

1.4 Demand

In ITSUMO, demand is basically represented by the population of drivers that use the network. Specific mechanisms are considered for demand routing, deadlock handling, driver definition, and en-route replanning.

1.4.1 Routing of the Demand

Demands are normally represented by an OD (origin-destination) matrix that results from some survey or other kind of measurement of demand. ITSUMO can also generate synthetic demands (see "OD Matrix" in Figure 1.2). This can be done assigning either uniform probabilities to all nodes or to a set of selected nodes, or specific origin and destination probabilities to selected nodes. In the latter, node A for instance may originate 20% of the trips and collect 5% of them, whereas node B originates, say, 2% and collects 30%.

For each trip, a vehicle is generated and a route is assigned. This is in sharp contrast to the basic Na-Sch model where vehicles are treated as individual particles *without* a route. Rather, they are routed at each intersection with a probability to turn left, right, or continue in the same street.

So far ITSUMO has basically allowed the creation of vehicles as Na-Sch particles, or provided a GUI to define route for only a handful of routes, those that were assigned to the so-called "floating cars" (FC). This process of route definition was manual and could not be done for many vehicles.

With this new version, to generate routes for vehicles, ITSUMO can use various algorithms as shown in the third block of Table 1.1. Besides the well-known Dijkstra and A*, ARA* stands for Anytime Repairing A* (Likhachev et al., 2008), a heuristic search algorithm. We have also implemented a dynamic shortest-path algorithm that uses dynamically changing quantities (e.g., traffic volume) as links' weights. This is normally used for replanning.

No matter whether the algorithm is used, the routing can be done either in a centralized way (e.g., routes are computed in a centralized manner and are assigned to vehicles), or in a decentralized way. The centralized case is trivial and is performed as in commercial simulators: given an OD matrix, an algorithm computes routes for each driver, simulates the journeys, and performs further reassignments until an equilibrium is found.

In the decentralized case, there is a decoupling vehicle driver because this allows the specification of several classes of drivers' behavior ("Drivers" in Figure 1.2). Here, the driver computes its own route based on a given strategy and on local knowledge. Therefore we refer to this as planning and discuss it in Section 1.4.3.

1.4.2 Deadlock Handling

In the previous version of ITSUMO, deadlocks were sometimes observed. The reason for their occurrence is that vehicles block each other and none is able to move further. In order to investigate the reasons and effects of this, in (Bazzan et al., 2011) a detailed study was conducted and one of the conclusions is that deadlocks are more harmful in the case of regular networks. In a previous version of the present chapter (Bazzan et al., 2010a) (where the case study was not a regular network), we have presented results that do not treat this type of deadlock.

In the present chapter we report this new extension, as well as results obtained using the deadlock handling mechanism. This mechanism works as follows. When there is no space for inserting a vehicle in the next link of its route, this vehicle is temporarily removed from the simulation. Its reinsertion occurs when two conditions are satisfied. First, the next link in its route is no longer full. Second, its current link must not be completely full either. Currently we consider that a vehicle may return to its current link if this is below 80% of its capacity. The time the vehicle remains out of the simulation does count in its travel time (because if not removed the vehicle would be blocked anyway). As mentioned, such a mechanism was used in the experiments reported in Section 1.5.

1.4.3 Driver Definition

Modeling drivers' behavior can be approached in different ways, depending on the purpose of the simulation. In some cases, the objective is to simulate the collective or macroscopic behavior. However, this behavior emerges out of individual ones. Simple algorithms, like the CA model, can be used to describe the movement of vehicles. However, this model does not provide support for modeling more sophisticated driver behavior such as that based on route planning or en-route decisions, which are appealing to AI practitioners (see Bazzan et al., (2011) for more details).

Next, the recent extensions that were made to ITSUMO in order to allow the definition of classes of drivers are discussed. This discussion concentrates here on two aspects: the prejourney planning and the en-route (re)planning.

The centralized assignment discussed in the previous section works only to the extent that the full rationality assumption is considered: drivers want to maximize their individual utilities. This econometric model considers neither bounded rationality nor individual preferences. However it is a relatively strong assumption that drivers know the whole traffic network, much less the traffic status of each link at each moment. The first assumption can only be accepted if all drivers are known to have GPS-based devices to guide them. The second (accurate, instantaneous knowledge of the status of each link) is still far from reality. Even when this is deployed, it is questionable whether drivers can indeed process all the information. Therefore, the centralized routing only addresses macroscopic investigations, for example, those that are carried out for urban planning purposes. It is not efficient for modeling real behavior of individual drivers though.

In the decentralized route computation, it is assumed that the driver itself will plan its journey, given its (partial) knowledge of both the traffic network and of the current traffic status. ITSUMO has a series of methods to allow the implementation of routing at the driver level. So far, without need of further coding, it is possible to use the algorithms that are mentioned in Table 1.1, both in centralized and decentralized variants. Of course, in the latter links' weights may differ from driver to driver as they are local perceptions.

1.4.4 Drivers and En-Route Replanning

One of the features of an autonomous driver is its ability to replan during the trip when facing congestion. In classical centralized approaches, this is hard to do due to the fact that each driver may have its own replanning strategy, own knowledge about traffic status, as well as own preferences and idiosyncrasies. In order to facilitate these definitions, ITSUMO allows each class of driver to have its own profile.

In Table 1.1 (last block) we show the possibilities for *en-route* replanning. One possibility is the trivial Na-Sch rerouting of vehicles but here there is actually no planning because these vehicles are treated as particles that are randomly rerouted.

The actual replanning, the one that happens autonomously at driver's level is based on a driver's perception of the congestion level. So far we assume that drivers only have local perception, and hence partial knowledge of the traffic status. However, the simulator is prepared to deal with situations in which drivers have full knowledge.

En-route replanning can be done using one of the algorithms mentioned. In all cases, a driver will compute a new route from the point where he/she starts to replan to the destination. If dynamic shortest-path algorithms are used, then the current traffic status of the known links is used. For unknown links, the length is used instead.

This means that when a driver arrives at a link $e^i \in P^j$, where P^j is the initially computed route of vehicle j, he/she evaluates how delayed he/she is when compared to the expected time. If the current time step is τ times higher than the expected time step, then the driver replans the route. Besides, the exact portion δ of the route where the driver considers replanning is also configurable. More anxious drivers will start replanning sooner. Therefore δ and τ (among other factors) define different classes of drivers. At the end of the simulation, one can evaluate the performance of each class of driver and compare these.

1.5 Case-Study: Aggregating Intelligence to Traffic Simulation

In order to illustrate the use of ITSUMO with new facilities for demand handling, we discuss a case study. This refers to the city of Porto Alegre (3005'S, 5110'W) in Brazil (Figure 1.5). The downtown part of the city was selected and exported from OSM and parsed to ITSUMO format using "OSM2ITSUMO." We discuss scenarios with and without traffic lights. When

Table 1.2: Average travel times (time steps): comparison.

Vehicles μ	Fixed Time			Greedy			IGMN	
	σ	μ		σ	μ		σ	
2000	313.1	1.7		312.7	3.6		309.0	4.0
2500	369.1	4.6		357.6	1.9		343.3	4.8
3000	388.5	5.3		375.0	4.6		358.5	4.6
3500	433.8	9.6		408.8	2.3		388.7	7.0
5000	479.1	7.0		433.9	3.7		415.1	6.5
7500	520.2	10.2		446.0	3.9		426.1	6.9

these are present, the signal plans are generated automatically using a cycle length of 60 s, with uniform green time for all phases. Thus, if an intersection has only two phases, each receives 30 s of green time. Each intersection has a traffic light agent that runs a control method. Here we discuss the following ones:

- Fixed time: the split of 30–30 s is not changed during the simulation;
- greedy strategy: changes the basic 30–30 split, increasing the green time for the most congested approach (and decreasing by the same proportion for the other approach);
- agents implement the IGMN approach discussed in Section 1.3.2.

Overall, that area of the city comprises 159 nodes (96 having traffic lights), 225 links totalizing 39 km. We remark that since each cell has 5 m, the network holds up to approximately 8K vehicles. The number of vehicles was varied from 2K to 7.5K, an increase regarding the scenario presented and discussed in (Bazzan et al., 2010a). However, this does not mean that all of them occupy the network at the same time as some may arrive earlier than others at their respective destination. For the purpose of illustration we have simply assumed that the demand is uniformly distributed among all nodes.

Other simulation parameters are as follows: unless mentioned, the number of simulation steps is 6000; the cell size is 5 m; and the maximum speed is 3 cells per simulation step (roughly 54 km/h if one-time step corresponds to 1 s).

We discuss next a simple example where drivers plan their routes using the Dijkstra algorithm taking the length of the link as weight. Although the simulator allows the output of the volume or occupancy of all links in order to generate histograms of links load, here we present the average travel time (for all drivers), for the situations without traffic lights, with traffic lights and fixed time plans, with greedy control, and with the IGMN method introduced in Section 1.3.2. In all cases, we present the average and deviation over 10 repetitions of the same setting. Travel time is measured in time steps. To be consistent with the before mentioned cell size of 5 m, it is assumed that one-time step equals 1 s. These results appear in Table 1.2.

For the greedy and IGMN, the following setting was used. The state of an agent is given by the states of the approaching links (number of stopped vehicles in each laneset) averaged using a time window corresponding to 300 s of simulated traffic. In particular, for IGMN, since there are two one-way approaching links for each traffic light, the possible state-space for each agent is a two-dimensional continuous plan. Thus, instead of using a rough discretization of states (e.g., low, medium, and high occupancy, as is the case in most RL-based works), each agent uses the actual average number of vehicles waiting at each laneset. As mentioned, this continuous state results in dimensionality problems related to the state space, thus requiring a function approximator such as IGMN to deal with continuous information.

The reward for each agent using IGMN is computed locally using the links' capacity, that is, it corresponds to the inverse of the square of the number of vehicles in the incoming lanesets. Regarding actions, these consist of a continuous value between 5 and 55, which corresponds to the green time of each phase. Each agent senses the environment each 20 s and may act after 600 s of simulated traffic. Finally, the value used for the IGMN parameters are: $\alpha = 0.1$; $\gamma = 0.7$; $\delta = 0.01$; and $\varepsilon = 0.1$ (see Heinen and Engel, 2010).

Analyzing Table 1.2, one notices that more sophisticated control strategies such as the IGMN do pay off when the number of vehicles is high, especially when the occupancy is above 50% of the network. Close to the saturation level, the gain in travel time represents 18% (IGMN over fixed time), or 4% (IGMN over greedy).

This simple, yet valid, scenario was given just for the sake of illustration of the use of the simulator, and to show what kind of results one may extract. It is not the purpose here to make comparisons with other tools, because most of them do not allow, for instance, implementation of more sophisticated control strategies such as IGMN, or en-route replanning.

To give an idea of scalability, we remark that we have also run simulations that cover the whole city of Porto Alegre. In this case we did not include all intersections and all streets, just the main arterials. These arterials however can hold around 100K drivers. Such results can be found in (Bazzan et al., 2011).

1.6 Conclusion

This chapter has discussed some recent extensions implemented in ITSUMO, an open-source microscopic traffic simulator that allows modeling of individual drivers or classes of drivers along with the implementation of different traffic-light control strategies. We plan to extend ITSUMO to consider other kinds of information such as that related to V2V communication and information provided via Internet and/or mobile phone in order to compare the performance of informed versus noninformed drivers.

Acknowledgment

This project and the authors are partially funded by CNPq and FAPERGS.

References

Balmer, M., Meister, K., Rieser, M., Nagel, K., Axhausen, K.W., 2008. Agent-based simulation of travel demand: structure and computational performance of MATSim-T. 2nd TRB Conference on Innovations in Travel Modeling. Portland.

Bazzan, Ana L.C., do Amarante, Maicon de B., Azzi, Guilherme G., Benavides, Alexander J., Buriol, Luciana S., Moura, Leonardo, Ritt, Marcus P., Sommer, Tiago, June 2011. Extending traffic simulation based on cellular automata: from particles to autonomous agents. Burczynski, Tadeusz, Kolodziej, Joanna, Byrski, Aleksander, Carvalho, Marco (Eds.), Proceedings of the Agent-Based Simulation (ABS) 2011, vol. 1, ECMS, Krakow, pp. 91–97.

Bazzan, A.L.C., Maicon de Brito do Amarante, Sommer, T., Benavides, A.J., ITSUMO, 2010. An agent-based simulator for ITS applications. In: Rossetti, R., Liu, H., Tang, S. (Eds), Proceeding of the 4th Workshop on Artificial Transportation Systems and Simulation. IEEE.

Bazzan, A.L.C., de Oliveira, D., da Silva, B.C., 2010b. Learning in groups of traffic signals. Eng. Appl. Artif. Intel. 23, 560–568.

Bazzan, Ana L.C., Wahle, J., Klügl, F., 1999. Agents in traffic modeling - from reactive to social behavior. In Advances in Artificial Intelligence, number 1701 in Lecture Notes in Artificial Intelligence, pages 303-306, Berlin/Heidelberg. Springer. Extended version appeared in Proc. of the U.K. Special Interest Group on Multi-Agent Systems (UKMAS), Bristol, UK.

Bonabeau, E., Theraulaz, G., Dorigo, M., 1999. Swarm Intelligence: From Natural to Artificial Systems. Oxford University Press, New York, USA.

Burmeister, B., Doormann, J., Matylis, G., 1997. Agent-oriented traffic simulation. T. Soc. Comp. Simul. 14 (2), 79–86.

Kurt Dresner and Peter Stone. Multiagent traffic management: a reservation-based intersection control mechanism. In: N.R., Jennings, C., Sierra, L., Sonenberg, M., Tambe. (Eds) Proc. of the International Joint Conference on Autonomous Agents and Multi-Agent Systems. pp. 530–537. New York, USA. July, 2004. IEEE Computer Society.

Heinen, M.R., 2011. A connectionist approach for incremental function approximation and on-line tasks. Ph.D. Thesis. Informatics Institute – Universidade Federal do Rio Grande do Sul (UFRGS). Porto Alegre, RS, Brazil. March, 2011.

Heinen, M.R., Engel, P.M., 2010. An incremental probabilistic neural network for regression and reinforcement learning tasks. In: Proceeding 20th International Conference Artificial Neural Networks (ICANN 2010), 6353 of LNCS, pp. 170–179. Thessaloniki, Greece. September, 2010. Springer-Verlag. 18 ITSUMO: an Agent-Based Simulator for ITS.

Likhachev, Maxim, Ferguson, Dave, Gordon, Geoff, Stentz, Anthony, Thrun, Sebastian, 2008. Anytime search in dynamic graphs. Artif. Intel. 172 (14), 1613–1643.

Nagel, K., Schreckenberg, M., 1992. A cellular automaton model for freeway traffic. Journal de Physique I 2, 2221.

de Oliveira, D., Bazzan, A.L.C., 2006. Emergence of traffic lights synchronization. In: Zobel, R., Borutzky, W., Orsoni, A. (Eds), Proceedings of the 20th European Conference on Modeling and Simulation (ECMS 2006, ABS Track), 572–577. ECMS, May 2006.

de Oliveira, Denise, Bazzan, Ana L.C., 2009. Multi-agent learning on traffic lights control: effects of using shared information. In: Bazzan, Ana L.C., Kluegl, Franziska (Eds.), Multi-Agent Systems for Traffic and Transportation. IGI Global, Hershey, PA, pp. 307–321.

de Oliveira, D., Bazzan, A.L.C, da Silva, B.C., Basso, E.W., Nunes, L., Rossetti, et al., 2006. Reinforcement learning based control of traffic lights in non-stationary environments: a case study in a microscopic simulator. In: Dunin-Keplicz, B., Omicini, A., Padget, J. (Eds). Proceedings of the 4th European Workshop on Multi-Agent Systems, (EUMAS06), 31–42.

Rossetti, Rosaldo, Liu, Ronghui, 2005. A dynamic network simulation model based on multi-agent systems. In: Klügl, F., Bazzan, A.L.C., Ossowski, S. (Eds.), Applications of Agent Technology in Traffic and Transportation, Whitestein Series in Software Agent Technologies and Autonomic Computing. Basel, Birkhauser, pp. 181–192.

da Silva, B.C., Basso, E.W., Bazzan, A.L.C., Engel, P.M., 2006. Dealing with non-stationary environments using context detection. In: Cohen, W.W., Moore, A., (Eds) Proceedings of the 23rd International Conference on Machine Learning ICML, pages 217–224. New York, ACM Press.

da Silva, B.C., Junges, R., Oliveira, D., Bazzan, A.L.C., 2006. ITSUMO: an intelligent transportation system for urban mobility. In: Nakashima, H., Wellman, M.P., Weiss, G., Stone, P., (Eds), Proceedings of the 5th International Joint Conference on Autonomous Agents and Multi-agent Systems, AAMAS, ACM Press. pp. 1471–1472.

Tumer, K., Welch, Z.T., Agogino, A., 2008. Aligning social welfare and agent preferences to alleviate traffic congestion. In: Padgham, L., Parkes, D., Mueller, J., Parsons, S. (Eds), Proceedings of the 7th International Conference on Autonomous Agents and Multi-agent Systems, IFAAMAS, Estoril. pp. 655–662,

van Katwijk, R.T., van Koningsbruggen, P., De Schutter, B., Hellendoorn, J., 2005. A test bed for multi-agent control systems in road traffic management. In: Kluegl, F., Bazzan, A.L.C., Ossowski, S. (Eds.), Applications of Agent Technology in Traffic and Transportation, Whitestein Series in Software Agent Technologies and Autonomic Computing,. Basel, Birkhauser, pp. 113–131.

Vasirani, Matteo, Ossowski, Sascha, April 2009. Exploring the potential of multi-agent learning for autonomous intersection control. In: Bazzan, A.L.C., Kluegl, F. (Eds.), Multi-Agent Systems for Traffic and Transportation. IGI Global, Hershey, PA, pp. 280–290.

Vasirani, Matteo, Ossowski, Sascha, June 2011. A computational market for distributed control of urban road traffic systems. IEEE Trans. Int. Transp. Systems 12 (2), 313–321.

Watkins, Christopher J.C.H., Dayan, Peter, 1992. Q-learning. Mach. Learn. 8 (3), 279–292.

CHAPTER 2

A Pattern-Based Framework for Building Self-Organizing Multi-Agent Systems

Manoel T. de Abreu Netto, Baldoino F. dos Santos Neto, Carlos J.P. de Lucena
Department of Computer Science, PUC-Rio, Rio de Janeiro, RJ, Brazil

2.1 Introduction

The approach of self-organizing systems has increased its relevance and is used to deal with complex domains. The use of this approach enables the development of decentralized systems that exhibit certain dynamicity and adaptability to couple with previously unknown perturbations (Serugendo et al., 2003, 2005). According to principles of self-organization, each component of the autonomic system obtains and maintains only local information available in the environment, in a decentralized way and without any external control, being restricted only to local interactions. It is on the basis of these interactions that the system exhibits it macroscopic behavior, which may be observed from a global point of view (Huebscher and McCann, 2008).

The multiagent system paradigm has been considered a promising solution for the building of self-organized systems (Sycara, 1998; Jennings and Wooldridge, 2000). Multiagent systems are a group of agents that cooperate to autonomously achieve a required systemwide or macroscopic behavior by using only local interactions, local activities of the agents, and locally obtained information.

Although the multiagent system paradigm is a feasible solution for the construction of self-organized systems, two of the main challenges faced today are: (1) the lack of mechanisms to regulate the interaction and coordination among the agents in the environment and (2) the infrastructure necessary to implement such mechanisms. Several works, (Gardelli et al., 2007; De Wolf and Holvoet, 2006; Babaoglu et al., 2006; De Wolf, 2007) have provided guidelines for expressing a series of recurrent design patterns present in self-organizing systems as a first step to face the lack of mechanisms to regulate agent interaction and coordination. Self-organizing patterns provide the directions and motivate the reuse of solutions already adopted in the process of constructing such systems. However, there is currently no infrastructure to support the implementation of the self-organizing multiagent system based on the proposed patterns.

In this chapter we propose JASOF – Jadex Self-Organization Framework, a pattern and a BDI-based framework that extends Jadex (Poukahr and Braubach, 2007) by providing a set

Advances in Artificial Transportation Systems and Simulation.
Copyright © 2015 Zhejiang University Press Co., Ltd. Published by Elsevier Inc. All rights reserved.

of capabilities, composed of plans and goals that implement the basic patterns described in (Gardelli et al., 2007). The primary goal of JASOF is to contribute to the self-organization area through software engineering techniques. This is accomplished by providing a framework that allows for implementation mechanisms to regulate the interaction and coordination between agents of a self-organized system through the reuse of previously proposed solutions (Sudeikat and Renz, 2008; Sudeikat et al., 2009; Fayad and Johnson, 1999; Neto et al., 2009a; Neto et al., 2009b).

The chapter is organized as follows. In Section 1.2, JASOF is detailed and Section 1.3 reports a case study while describing how to extend the framework to implement a solution to the automatic guided vehicle problem. Section 1.4 presents some related work. Finally, Section 1.5 concludes this study and presents some future work.

2.2 JASOF

In this section, we describe the main idea of the proposed framework, followed by a description of how to implement the environment and patterns available to regulate and coordinate the agents that act in such an environment.

2.2.1 Main Idea

Due to the complexity of building self-organizing systems, (Gardelli et al., 2007; De Wolf and Holvoet, 2006; Dobson et al., 2006) have elaborated a series of recurrent design patterns that are commonly present in these systems. The aforementioned patterns aim to smooth the process of constructing self-organizing systems by offering a set of directions and making adequate reuse of solutions that have already been tested and documented.

In this context, the JASOF framework, which is based on the BDI (Weiss, 1999) paradigm extending Jadex, provides the necessary infrastructure to implement the patterns and the environment where the agents operate. In Jadex, from the external point of view, the agent is a black box that receives and sends messages. Messages, internal events, and new goals are used as input for the reaction and deliberation mechanism (*Reaction Deliberation*). Then, based on these results of deliberation, the mechanism directs the events to the plans that are already running or those that can be called up from the library plans. The plans, when running, can modify the knowledge base of the agents, send messages to other agents, create new subtargets, and enable internal events. The reaction mechanism and resolution is generally the same for all agents. However, the behavior of a specific agent is determined by its beliefs (*Beliefs*), goals (*Desires*) and plans (*Intentions*). The patterns in JASOF are provided as a set of goals and plans encapsulated as Jadex capabilities, which can be easily imported and implemented by an agent.

The environment is modeled as a space composed of interconnected positions following the notion of a grid. Furthermore, it is based on the agents-and-artifacts (A&A) approach (Gardelli et al., 2008), which considers the environment populated by agents and artifacts,

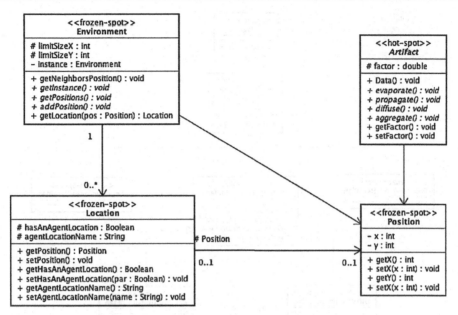

Figure 2.1: JASOF Class Diagram (Environment)

where artifacts are reactive agents providing resources and services, meaning an active environment. The class diagram in Figure 2.1 shows the structure of the environment, where the class *Environment* contains information about the available locations, the agents in the system and your neighborhood. These locations are represented by the class *Location* that provides information about its artifacts (class *Artifact*), their manager agent and their relative position in the environment (class *Position*).

2.2.2 JASOF Structure

First, we show how the environment is modeled and how it acts in our framework through the *Agent Location*. Then, the plans provided by JASOF that enable the implementation of the referred patterns are described in detail. Figure 2.2 shows the class diagram that represents these plans and the hotspots from which it is possible to create new patterns not yet documented in the literature. In Figure 2.3 we have the basic JASOF architecture, showing the components of JASOF, described below, and the interconnection between the application and Jadex.

2.2.2.1 Environment

In natural self-organizing systems and artificial prototypes, the environment plays a key role in the system dynamics. It is through the interaction with the environment that self-organization of different systems is achieved. A typical example is that of ants foraging, which perform interactions with the environment such as storage of pheromones and their perception, which then guide their decisions and do their work. Therefore, the environmental model is a key

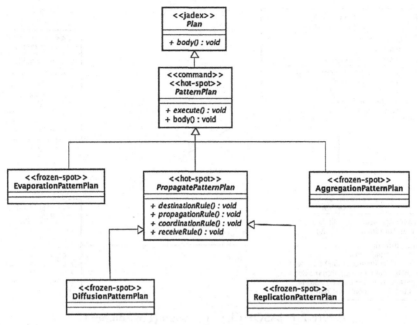

Figure 2.2: JASOF Class Diagram (Plans)

part of the framework to build such systems. It is from the environment that the patterns of self-organization are possible. Thus, it is available in the JASOF modeled as a set of discrete locations, interconnected to form a grid in two dimensions and, in addition, as mentioned above, is based on the A&A approach, giving it an active environmental characteristic. At

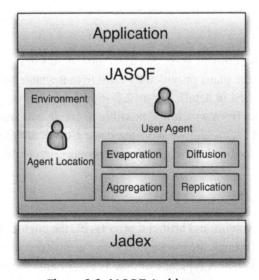

Figure 2.3: JASOF Architecture

each position of the environment a location is allocated. It is important to define the difference between them: position refers to a reference point in the environmental space, whereas location is an area located in a position that can be inhabited by multiple agents. In each location there is an agent responsible for its management; this agent is named *Agent Location*.

2.2.2.2 Agent location

As already said, each location in the environment has an agent, called *Agent Location*, responsible for managing the artifacts stored in this location, to execute the patterns and to provide artifact access to agents that are in the same position. It is through the agent *Location* that the environment has an active behavior, executing and interacting with the other participants of the system. Those participants are comprised of all agents in the environment, *Locations* and others. The agents that are not *Location* are called *User* agents in the framework.

Through communication with the *Agent Location* of a specific location it is possible for a given agent to insert an artifact, which can represent relevant information such as a pheromone, or to read the stored artifact. All communication with the *Agent Location* to insert and read is a frozen spot of the framework. The reason for associating agents to each location is justified by the need to search for forms of decentralization, despite the large number of agents and the communication required for reading and writing information. In the following sections the ease the environment and agents have in implementing self-organization patterns can be observed.

2.2.2.3 Diffusion pattern

The idea behind the *Diffusion Pattern* is the process present in nature that when a pheromone is deposited into the environment, it spontaneously tends to diffuse to neighboring locations (Weyns et al., 2008). Diffusion distributes the information equally across all neighboring locations.

In order to implement the diffusion pattern it is necessary for the agent to import the *DiffusionPattern* capability, depicted in Figure 2.4. It provides the infrastructure needed for an agent to send information, decreased by a factor, to a set of neighbors selected according to a rule (*PropagationRule*) and perform the actions related to the coordination process or react to a received message. This is, in essence, the coordination mechanism. Such capability has the class *DiffusionPatternPlan*, Figure 2.2, which needs to be extended in order to define the rules for selection of agents to receive the message (method *DestinationRule*), the actions of a coordination (method *CoordinationRule*) and the propagation (method *PropagationRule*) process. As we can see in Figure 2.4, the message *diffusionMsg* is the standard event used to perform the diffusion pattern, so importing that capability implies that the agent can receive and handle that message. If necessary it is possible to modify the *ReceiveRule* method and change this process.

```
<beliefs>
    <beliefsetref name="artefact" class="Artefact" exported="true">
        <abstract/>
    </beliefsetref>
    <!-- Locations Neighbors -->
    <beliefsetref name="neighbors" class="Position" exported="true">
        <abstract/>
    </beliefsetref>
</beliefs>
 <plans>
    <plan name="diffusion_plan">
        <body class="DiffusionPatternPlan"/>
        <waitqueue>
            <messageevent ref="diffusionMsg"/>
        </waitqueue>
    </plan>
</plans>
<events>
    <messageeventref name="diffusionMsg" exported="true">
        <abstract/>
    </messageeventref>
</events>
```

Figure 2.4: Diffusion Capability

2.2.2.4 Evaporation pattern

Evaporation can be considered a mechanism to reduce the amount of information, based on a time relevance criterion, avoiding an overload of information.

The *Evaporation Pattern* can be implemented by an agent, using JASOF, by importing the *EvaporationPattern* capability. Figure 2.5 shows that such capability will periodically execute

```
<beliefs>
    <beliefsetref name="artefact" class="Artefact" exported="true">
        <abstract/>
    </beliefsetref>
</beliefs>

<goals>
    <performgoal recur="true" recurdelay="6000" name="evaporation" exported="true">
        <unique/>
    </performgoal>
</goals>

<plans>
    <plan name="evaporation_plan">
        <body class="EvaporationPatternPlan"/>
        <trigger>
            <goal ref="evaporation"></goal>
        </trigger>
    </plan>
</plans>
```

Figure 2.5: Evaporation Capability

the *evaporate()* method, present in the class *Artifact*, of the instances contained in the *Belief Base* of the respective agent *Location*. To change the frequency it is necessary to modify the parameter *recurdelay* that by default is 6000, as shown in Figure 2.5. Another parameter that can be determined is how much of the information will be evaporated per cycle; in this case the attribute *factor* of the *Artifact* class must be changed.

2.2.2.5 Aggregation pattern

Aggregation is a mechanism of reinforcement and is also observable in human social tasks (Gardelli et al., 2007). When redundant information is stored, its relevance increases.

As in the other patterns, to implement the Aggregation patterns it is necessary to import the *EvaporationPattern* capability, which captures the concept of Evaporation, periodically checks the artifacts saved in the Belief Base and checks which of them may suffer an aggregation. This aggregation is accomplished by the method *aggregate()* of the *Artifact* class. An aggregation process between two artifacts produces one with more relevance.

2.2.2.6 Replication pattern

Natural systems typically feature replication mechanisms in order to increase security and robustness. The Replication pattern works in a way similar to the Diffusion pattern; the main difference resides in the factor of the information replicated. In the diffusion case, the artifact suffers a decrease in the factor attribute and, on the other hand, the replication does not modify that attribute, keeping its actual value. The mechanisms *DestinatioRule*, *CoordinationRule*, and *PropagationRule* are used for the same purpose.

2.2.3 JASOF Hotspots

The framework hotspots are:

1. The ability to specify the information to be used in the communication process: Using the class Artifact it is possible to create new types of data to be stored and exchanged by agents.
2. Pattern composition for the creation of new self-organizing mechanisms: Through Jadex features it is possible to include several plans in an agent, creating the potential for patterns composition.
3. The creation of new basic patterns: Flexibility through the extension of the PatternPlan class and the implementation of its respective plan.
4. Coordination and propagation mechanisms to be specified in each pattern: It is possible to build a self-organizing solution on the basis of ant foraging or in a Gossip Protocol, just modifying this mechanism.

2.3 Case Study: A Self-Organized Automatic Guided Vehicle

Automated guided vehicles (AGVs) are fully automated, custom-made vehicles that are able to transport incoming loads (i.e., packets, materials, and/or products) in a logistical or production factory environment (Weyns and Holvoet, 2008; Weyns et al., 2008). AGV systems can be used for distributing manufactured products to storage locations or as an inter-process system between various production machines. An AGV system receives transport requests for loads from a factory management system or machine operating software, and instructs AGVs to execute transports.

We have used the JASOF framework to build a self-organizing multiagent system in order to implement a decentralized solution to the AGV problem. Furthermore, the system can withstand perturbations, such as unavailable routes or barriers in the path.

2.3.1 Main Idea

In order to implement a solution to this problem we have modeled the AGV system using the JASOF framework with four agents: (1) Destination Agent, representing the destination; (2) Warehouse Agent, representing the warehouse; (3) Transporter Agent, representing the AGV; (4) Location Agent, representing the Stations between the warehouses and the destinations.

As we can see in Figure 2.6, which illustrates a reduced situation, the architecture is modeled as follows: the locations where packages must be picked up are located on the left (Warehouse), the destinations on the right (Destination), while the network in between consists of Locations and connecting segments on which AGVs are located (AGV Location). Each individual AGV (Transporter) is capable of only a limited set of local Actions, such as Move, Pick Load and Drop Load. The vehicles move loads in a factory by moving towards the warehouse location: picking up the load, moving towards the destination location, and dropping the load. The fleet of AGVs has to self-organize efficiently to accommodate a current request for transport and to withstand unpredictable situations. In the next subsection we present more details about such agents.

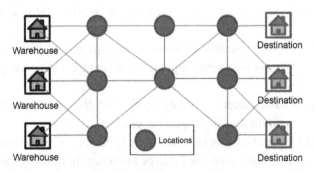

Figure 2.6: AGV Environment Structure

2.3.2 Destination Agent

In our solution using the framework, this agent is responsible for receiving the packages transported by the Transporter agent and to inform, using the diffusion pattern, the neighborhood of what it is capable of doing. This information is communicated to the Location agents that are interconnected by one segment, excluding the other types of agents.

This agent was implemented as a Jadex agent importing the *DiffusionPattern* capability of the framework. This pattern has a plan with four mechanisms of extensibility: receive, destination, coordination, and propagation. Receive and destination mechanisms were not modified though. The standard solution was adopted. However, the *DestinationRule* was extended, regulating that only agents Location were able to receive the *diffusionMsg*. Furthermore, the *PropagationRule* was extended to determine the type of information to be diffused. Complementing the functionality of the Destination agent, a new plan was created to define the protocol specifying the messages used to receive new packages from the Transporter agent.

2.3.3 Warehouse Agent

This agent is responsible for managing the presence of packages in its location. When a new package arrives, the agent runs the diffusion pattern to propagate the information in the environment, advising that a new load is available and ready to be transported. As the Destination agent, the mechanisms *ReceiveRule* and *CoordinationRule* have been maintained as the default option and the *DestinationRule* and *PropragationRule* were modified for the same purpose. The information diffused contains the position of the Destination agent in the environment, used by the agent Transporter to build its route.

2.3.4 Transporter Agent

This agent represents the AGV, responsible for transporting packages from warehouses to their destinations. Their actions are: Move, Pick Load and Drop Load. Each action was modeled as a step in a recurrent goal: while the Transporter is not loaded with a package to be transported, the agent will move (action Move) past its neighbors for information about locations that may eventually have a package to be transported; this movement is performed until the agent finds a package. In our solution the battery was not considered a problem. Once a package is found, the Transporter agent carries the load (Pick Load) and then the plan to deliver the package is triggered. The agent will travel through the environment looking for the destination and will build a route collecting the information available in the locations. When the agent finds the destination, a message is sent to the Destination agent at that position to dispatch the package. Finally, the process is restarted.

2.3.5 Location Agent

On this occasion, the Location agent performs the Evaporation, Aggregation, and Diffusion patterns. The Evaporation pattern, as seen in the Section 1.2.2.4, occurs periodically performing the decrement of the relevance factor from the information stored in the location. It is the Aggregation. However, in this case, aggregation is performed between similar information stored. The Diffusion pattern occurs when the *diffusionMsg* received from other agents has the parameter *path* higher than 0, indicating that the information has to be diffused again.

2.3.6 Execution

The execution can be described through the six events that occur. They are:

1. Initializing the environment; the locations are put in their positions and their respective Location agents are instantiated as well as the agents Warehouse and Destination. As we can see in Figure 2.7.
2. As soon as the Destination agent is created, the diffusion process starts by sending information to the neighborhood about its availability and position. As we can see in Figure 2.8.
3. Adding a package – when a package is loaded in a Warehouse, the responsible agent starts to diffuse to the neighborhood that a new package is waiting to be carried out. As shown in Figure 2.9.

```
DEBUG: new AGVEnvironment
DEBUG: Adding position 1x1
DEBUG: Adding position 1x2
...
DEBUG: Adding position 5x3
DEBUG: Creating Plan....
DEBUG: Adding location in position 1x1
DEBUG: Adding location in position 1x2
...
DEBUG: Adding location in position 5x3
...
DEBUG: Alocating Agent in position: 2x1
DEBUG: AgentLocation_1 Executing EvaporationPatternPlan...
DEBUG: Alocating Agent in position: 2x2
DEBUG: AgentLocation_2 Executing EvaporationPatternPlan...
...
DEBUG: AgentLocation_7 Executing EvaporationPatternPlan...
DEBUG: Alocating Agent in position: 4x3
DEBUG: AgentLocation_8 Executing EvaporationPatternPlan...
DEBUG: Alocating Agent in position: 1x1 (Warehouse)
DEBUG: Alocating Agent in position: 1x2 (Warehouse)
DEBUG: Alocating Agent in position: 1x3 (Warehouse)
DEBUG: Alocating Agent in position: 5x1 (Destination)
DEBUG: Alocating Agent in position: 5x2 (Destination)
```

Figure 2.7: Initializing the Environment

```
DEBUG: AgentDestination_12 Diffusing 'destination'...
DEBUG: Creating diffusionMsg...
DEBUG: Setting the content...
DEBUG: Getting neighbors...
DEBUG: 3 neighbors found...
DEBUG: Getting the environment...
DEBUG: Getting location of position: 5x2
DEBUG: No agent at this position...
DEBUG: Getting location of position: 4x1
DEBUG: Adding a receiver : AgentLocation_6@machine1...
DEBUG: Getting location of position: 4x2
DEBUG: Adding a receiver : AgentLocation_7@machine2..
DEBUG: Sending the message...
DEBUG: AgentLocation_7 diffusionMsg Receive...
DEBUG: AgentLocation_6 diffusionMsg Receive...
DEBUG: Data type: dpt size: 5
DEBUG: Saving the DISPATCH in the beliefbase...
```

Figure 2.8: Diffusion Process

4. While these events occur, the Location agents have already started the evaporation process. The ones near the agent Destination and Warehouse evaporate the recent information about position and package, respectively.

5. The Transporter agent is looking for information about available packages to be picked up. It is through the agents Locations that it builds its route, as we can see in Figure 2.10.

6. When the Transporter agent finds a package, it starts to look for information about destinations, similar to the step described above. When the destination is found the Transporter agent reinitializes the search for another package.

```
DEBUG: Executing UserAgentPlan ...
DEBUG: Transporter_16 Looking for a Warehouse...
DEBUG: Setting agentsId...
DEBUG: UserAgent Sending newDataDiffusionMsg to
    AgentWarehouse_1@sepultura
DEBUG: AgentWarehouse_1 received newDataDiffusionMsg from:
    AgentIdentifier(name=UserAgent@sepultura)
DEBUG: AgentWarehouse_1 Getting the content...
DEBUG: Data received: pck:
DEBUG: Saving in the beliefbase...
DEBUG: Creating a diffusionMsg...
DEBUG: Setting the content... pck:1:1:2:info
DEBUG: Getting neighbors...
DEBUG: 3 neighbors found...
DEBUG: Getting the environment...
DEBUG: Getting location of position: 2x1
DEBUG: Adding a receiver : AgentLocation_1@sepultura...
DEBUG: Getting location of position: 1x2
DEBUG: No agent at this position...
DEBUG: Getting location of position: 2x2
DEBUG: Adding a receiver : AgentLocation_2@sepultura...
DEBUG: Sending the message...
```

Figure 2.9: Warehouse diffusing a package availability

```
DEBUG: Transporter_16 looking for destination...
...
DEBUG: Transporter_16 Moved...
DEBUG: Transporter_16 new position: 3x2
DEBUG: AgentLocation_5 received a dispatchPackage request...
DEBUG: Transporter_16 reply received...
DEBUG: Data received: null
DEBUG: Transporter_16 trying to move to 2x2...
DEBUG: Transporter_16 Moved...
...
DEBUG: Transporter_16 new position: 4x2
DEBUG: AgentLocation_7 received a dispatchPackage request...
DEBUG: Data to be replyed: dpt:5:1:0.9:info
DEBUG: AgentLocation_7 sending reply...AgentIdentifier
   (name=Transporter_16@sepultura)
DEBUG: Transporter_16 reply received...
DEBUG: Data received: dpt:5:1:0.9:info
DEBUG: Try position: 5x1
DEBUG: Transporter_16 trying to move to 5x1...
DEBUG: Transporter_16 Moved...
DEBUG: Requesting AgentDestination_12 dispatch...
DEBUG: AgentDestination_12 Dispatch requested...
*DEBUG: Package dispatched at AgentDestination_12
```

Figure 2.10: Dispatch Process

2.3.7 System Composition

A relevant motivation for the separation of the mechanisms of self-organization in simple patterns is the resulting possibility of forming more sophisticated and undiscovered ones, from just a simple composition of patterns. This separation is one of the main challenges in engineering self-organizing systems, as discussed in (Parunak and Brueckner, 2011). As noted, the JASOF framework allows the construction of agents that perform specific mechanisms of self-organization and their combination. Also, as the case study illustrates, we can have agents with different mechanisms interacting in the system, and the result of this composition can be the emergence of known high-level mechanisms, like stigmergy or gradient fields, or even others not yet described in the literature.

The AGV case study was modeled on the basis of the construction of a system supported by indirect communication among the Transporters through the environment, using the notion of pheromones. We used the standard Evaporation, Diffusion, and Aggregation patterns in Location entities. However, our solution to the problem in finding AGV routes could be based on the gradient fields idea, where from these same patterns we could compose a new solution, just changing the hotspots points: mainly the propagation method and the Artifact class. Another possible self-organizing solution for the AGV problem, using the JASOF framework, could be through the Gossip protocol that has been shown to be robust in the presence of failures of individual nodes (Gavidia et al., 2006). As an example, it has been used in wireless mesh networks to propagate a news device, or to disseminate state information with the purpose of building hierarchies of clusters in wireless sensor networks (Iwanicki and van Steen, 2009).

2.4 Related Work

In (De Wolf and Holvoet, 2006), two of the main decentralized coordination mechanisms are described in a structured manner. They are the gradient field and market-based control. These mechanisms are dealt with as patterns to be adopted when searching for a decentralized organization solution. Furthermore, (De Wolf, 2007) describes how these patterns need to be adopted and the consequences of their adoption. Nevertheless, the pattern descriptions are only described at the conceptual level, only allowing the formulation of high-level solutions. The authors are not concerned about the implementation aspects, or in supporting the software development, that is, how to structure the self-organization system using agents and classes.

Following the same approach as De Wolf (De Wolf and Holvoet, 2006), (Gardelli et al., 2007) describes the above mentioned patterns in terms of low-level patterns, which can be called basic patterns. They are: Collective Sort, Diffusion, Aggregation, Replication, and Evaporation. However, the approach used is the same and there is still no concern with the implementation of these patterns.

2.5 Conclusions and Future Work

In this chapter we presented JASOF, a pattern-based framework that provides built-in self-organizing patterns implemented following the BDI architecture. Self-organizing patterns have only been described so far (Gardelli et al., 2007; De Wolf and Holvoet, 2006; Babaoglu et al., 2006) as a very high level of abstraction without any indication of how they could be implemented. In our proposed solution, self-organizing patterns modularized into capabilities are the basic building block to construct self-organizing multiagent systems or new self-organizing patterns.

The JASOF approach was illustrated using the AGV case study. The AGV encompasses three different agents: (1) Location — implements the evaporation, diffusions, and aggregation; (2) Warehouse – implements the Diffusion pattern; and (3) Destination – implements the Diffusion pattern. This case study shows the viability of JASOF for the implementation of complex and decentralized problems, in addition to illustrating how such problems can have a facilitated solution by the composition of self-organizing patterns and simply implemented using the BDI architecture or, more specifically, using agents goals and plans modularized into Jadex capabilities.

This framework is a first step towards a complete infrastructure where complex and decentralized systems can be automatically generated from high-level specifications, such as domain-specific languages. As a future work, we intend to validate our ideas in more complex case studies and suggest some guidelines to help the creation of new self-organizing patterns using our proposed infrastructure.

Acknowledgment

This work is supported by the Fundação de Amparo à Pesquisa do Estado do Rio de Janeiro – FAPERJ.

References

Babaoglu, O., Canright, G., Deutsch, A., Di Caro, G.A., Ducatelle, F., Gambardella, L.M., Ganguly, N., Jelasity, M., Montemanni, R., Montresor, A., Urnes, T., September 2006. Design patterns from biology for distributed computing. TAAS 1 (1), 26–66.

De Wolf, T., 2007. Analysing and engineering self-organising emergent applications. Ph.D. Thesis. Department of Computer Science, K.U.Leuven, Leuven, Belgium. May, 2007.

De Wolf, T., Holvoet, T., 2006. Decentralised coordination mechanisms as design patterns for self-organising emergent applications, Proceedings of the Fourth International Workshop on Engineering Self-Organising Applications. In: Brueckner, S., Hassas, S., Jelasity, M., Yamins, M. (Eds). Future University-Hakodate, Japan, pp. 40–61.

Dobson, S., Denazis, S., Fernández, A., Gaiti, D., Gelenbe, E., Massacci, F., Nixon, P., Saffre, F., Schmidt, N., Zambonelli, F., December 2006. A survey of autonomic communications. TAAS 1 (2), 223–259.

Fayad, M., Johnson, R., 1999. Building Application Frameworks: Object-Oriented Foundations of Framework Design (Hardcover), first ed. Wiley, ISBN-10: 0471248754.

Gardelli, L., Viroli, M., Casadei, M., Omicini, A., 2008. Designing self-organising environments with agents and artefacts: a simulation-driven approach. IJAOSE 2 (2), 171–195.

Gardelli, L., Viroli, M., Omicini, A., 2007. Design patterns for self-organizing multi-agent systems. In: Wolf, T.D., Sare, F., Anthony, R., (Eds), 2nd International Workshop on Engineering Emergence in Decentralised Autonomic System (EEDAS) 2007. ICAC 2007, Jacksonville, Florida, USA. June 2007. CMS Press, University of Greenwich, London, UK. pp. 62–71

Gavidia, D., Voulgaris, S., van Steen, M., 2006. A gossip-based distributed news service for wireless mesh networks. In: Proceedings. Third IEEE Conference on Wireless On demand Network Systems and Services (WONS). Les Menuires, France.

Huebscher, M.C., McCann, J.A., 2008. A survey of Autonomic Computing – Degrees, Models, and Applications. ACM Computing Survey.

Iwanicki, K., van Steen, M., 2009. Multi-hop cluster hierarchy maintenance in wireless sensor networks: a case for gossip-based protocols. In: Proceedings Sixth European Conference on Wireless Sensor Networks (EWSN). Cork, Ireland.

Jennings, N.R., Wooldridge, M., 2000. Agent-oriented software engineering. In: Bradshaw, J. (Ed.), Handbook of Agent Technology. AAAI/MIT Press.

Neto, B., Costa, A., Netto, M. T. A., Silva, V., Lucena, C., 2009a. JAAF: A Framework to Implement Self-Adaptive Agents. Proceedings of the 21st International Conference on Software Engineering and Knowledge Engineering. (SEKE'2009).

Neto, B., Costa, A., Silva, V., Lucena, C., Netto, M. T. A., 2009b. JAAF-S: A Framework to Implement Autonomic Agents Able to Deal with Web Services. Proceedings of the 4th International Conference Proceedings of the 4th International Conference on Software and Data Technologies (ICSOFT'09).

Parunak, H.V.D., Brueckner, S.A., May, 2011. Software engineering for self-organizing systems. In: Proceedings of Twelfth International Workshop on Agent-Oriented Software Engineering (AOSE 2011). Taipei, Taiwan.

Poukahr, A., Braubach, L., 2007. Jadex User Guide, Distributed System Group University of Hamburg, Germany, Release 0.96. <http://vsis-www.informatik.uni-hamburg.de/projects/jadex/>. June, 2007.

Serugendo, G. Di, M., Foukia, N., Hassas, S., Karageorgos, A., Mostéfaoui, S.K., Rana, O.F., Ulieru, M., Valckenaers, P., and Aart, C.V., 2003. Self-organisation: Paradigms and applications. In: Serugendo, G.Di.M., Karageorgos, A., Rana, O.F., Zambonelli, F., (Eds), Engineering Self-Organising Systems: Nature-Inspired Approaches to Software Engineering. LNCS (LNAI), 2977, pp. 1–19. Springer, May 2004. International Workshop on Engineering Self-Organizing Applications (ESOA 2003). Melbourne, Australia, July 2003.

Serugendo, G., Di, M., Gleizes, M.-P., Karageorgos, A., 2005. Self-Organisation in MAS. Knowledge Engineering Review 20 (2), 165–189, Cambridge University Press.

Sudeikat, J., Braubach, L., Pokahr, A., Renz, W., Lamersdorf, W., 2009. Systematically engineering self-organizing systems: The SodekoVS Approach. Proceedings des Workshops über Selbstorganisierende, adaptive, kontextsensitive verteilte Systeme (KIVS 2009).

Sudeikat, J., Renz, W., 2008. On the encapsulation and reuse of decentralized coordination mechanisms: A layered architecture and design implications. In: Communications of SIWN, vol. 4, no. ISSN 1757–4439, pp. 140–146.

Sycara, K., 1998. Multi-agent Systems. Artificial Intelligence 10 (2), 79–93.

Weiss, G., 1999. Multiagent Systems: A Modern Approach to Distributed Artificial Intelligence. The MIT Press, Cambridge, MA, USA.

Weyns, D., Holvoet, T., Jan. 2008. 2008. Architectural design of a situated multi-agent system for controlling automatic guided vehicles. Int. J. Agent-Oriented Softw. Eng 2 (1), 90–128 http://dx.doi.org/10.1504/IJAOSE.2008.016801.

Weyns, D., Holvoet, T., Schelfthout, K., Wielemans, J., 2008. Decentralized control of automatic guided vehicles: applying multi-agent systems in practice. In: Companion To the 23rd ACM SIGPLAN Conference on Object-Oriented Programming Systems Languages and Applications (Nashville, TN, USA, October 19–23, 2008). OOPSLA Companion '08. ACM, New York, NY, 663–674. doi:10.1145/1449814.1449819.

CHAPTER 3

An Agent Methodology for Processes, the Environment, and Services

Lúcio Sanchez Passos*, Rosaldo J.F. Rossetti*, Joaquim Gabriel**

*LIACC, DEI, Faculdade de Engenharia da Universidade do Porto, Porto, Portugal; **IDMEC, DEMec, Faculdade de Engenharia da Universidade do Porto, Porto, Portugal

3.1 Introduction

The growth of urbanization in the last 50 years has aggravated the transport issues in consequence of the high number of vehicles in urban areas. Moreover, two aspects must be emphasized: ecosystem impacts and deaths in traffic. In respect to the global concern with ecosystems, urban transport is regarded as a "villain" due to the fact that it is one of the main contributors to CO_2 emissions (U.S. Environment Protection Agency, 2009), accelerating global warming. Additionally, developing countries are experiencing a growth in deaths caused by vehicle accidents. For instance, in Brazil and in China, deaths in traffic exceeded deaths caused by weapons (International Transport Forum, 2009).

Observing this scenario, the lack of efficiency and the need for a remodeling of the transportation systems becomes evident. Road network expansion is not the unique solution; now, it is possible to use all new resources, namely communication networks, artificial intelligence, and others, to improve the system efficiency and safety. This new concept has already been presented within the Intelligent Transportation Systems (ITS) scope, and has been widely studied by the scientific community to fulfill the idea earlier.

What is the real scenario expected for the future? What are all the features and technologies that can help us to achieve these goals? It is still too hard to answer all these questions. Nevertheless, we can identify some main features of tomorrow's intelligent transportation. The system must take decisions automatically, perceiving the environment's conditions and coordinate its actions; also it demands some flexibility from the user's perspective so it can provide personalized services. Additionally, infrastructure should evolve to a stage where intelligence is inherent. As transport systems are rather spread out, Future Urban Transportation (FUT) requires a distributed architecture and advanced communication technologies to have the previously requirement fulfilled, including accuracy as an important aspect to reduce system failure situations as well.

Advances in Artificial Transportation Systems and Simulation.
Copyright © 2015 Zhejiang University Press Co., Ltd. Published by Elsevier Inc. All rights reserved.

Urban transportation systems are considered to be complex because they involve a large number of entities that act autonomously, a large number of influencing factors, for example, weather and ecosystems. All relations between entities and influencing factors are nonlinear, dynamic, and hard to model precisely. Thus, some experimentation problems emerge (Wang and Tang, 2005) that are extremely difficult to tackle, and many times infeasible to test in a real situation due to their complexity. From this discussion arose the concept of Artificial Transportation Systems (ATS), that is, the research area whose goal is to use the artificial society approach to validate and evaluate complex systems with computational experiments. Thus, it creates a virtual environment that makes possible an in-the-loop hardware/software interaction and elicitation techniques. Hence, the demand for a test platform to experiment new approaches (which permit FUT deployment), is fulfilled.

The multiagent systems (MAS) paradigm is undoubtedly the best metaphor to represent and overcome all presented issues in an ATS framework. According to Wang and Tang (2005), there is no technique that models all required aspects of complex systems. Nevertheless, the agent-oriented modeling methodologies can give a starting point, even with simple diagrams of participants, relations, and interactions, to properly represent such a domain.

An agent-oriented modeling methodology is proposed, focusing on complex scenarios, such as artificial transportation systems due to the lack of appropriate tools for this niche of applications. To do this we use the SME approach based on Gaia (Zambonelli et al., 2003), relating Gaia to service-oriented perspectives for its flexibility and for being user-centered. Another reason is that the business process paradigm is being used by analysts and designers to easily model complex behaviors of the environment in most MAS applications. To illustrate our approach, we instantiate a simple scenario and give guidelines that can be of help to the reader in future modeling tasks.

Following this brief motivation, the remaining of this chapter is organized as follows. In the next section, we discuss background concepts used to conceptualize our approach, such as agent-oriented methodologies focusing on Gaia and Tropos (Giunchiglia et al., 2002). The chapter's main section gives more detail where the methodology is presented and discussed, starting with the elicitation of requirements and, furthermore, analysis and design phases. We apply the approach to a transport scenario to instantiate it and to help us to better understand its behavior. Finally, conclusions are drawn and suggestions for future work are presented.

3.2 Background

Like any software, implementing agents must have an engineering process behind them which is generally called Agent-Oriented Software Engineering (AOSE). It aims to understand the features that an agent-based approach can bring to the deployment systems taking advantage of autonomy, heterogeneity and dynamism. Several AOSE methodologies have been proposed over the years (see Cossentino et al., 2010; Akbari, 2010 for surveys), deriving from traditional perspectives (such as object oriented), to innovative assumptions (e.g., swarming systems). Each methodology suggests a different development process covering some or all phases from the software engineering perspective. Due to the number of methodologies, the approach called Situational Method Engineering (SME) (Kumar and Welke, 1992) aims to support the reuse of existing ones and, at the same time, to enable customization of a specific methodology, on the basis of specific scenarios.

By the necessity to cover all phases of software engineering Tropos was created. The methodology uses, through all steps of development, the agent notion and related concepts to better address specifications. Besides, the early requirements analysis is a crucial part of capturing all goals. In addition, it adopts Eric Yu's i* model that provides notions of actors, goals and actors dependencies to model an application. A strong feature of Tropos is the requirements analysis which helps to group important information about system needs. On the other hand, it models the environment but does not give methods to design an actor who could be dynamic and interactive.

Another AOSE methodology is Gaia, originally proposed in 2000. It sees a multi-agent system as a computational organization where each role is interpreted by an agent who cooperates with other agents to achieve a determined goal. Although Gaia was quite often used, it presented limitations and a new version was proposed in Gaia v.2 (Cernuzzi et al., 2004). Some extensions aimed at overcoming these issues, namely ROADMAP (Juan et al., 2002), use of AUML (Odell et al., 2001), and others (Gonzalez-Palacios and Luck, 2008; Castro and Oliveira, 2008). Castro and Oliveira (2008) propose the use of UML 2.0 to complement and replace some aspects of the original Gaia. Nevertheless, it has an interesting feature of creating a service model and also correlating it with the agent model.

Although Gaia's improvements try to fulfill existing drawbacks, there still are other facets that have to be observed. The first limitation is properly captured dynamic environment's aspects, which are essential to complex MAS. In addition, organization structures are modeled using a simple notation for difficult construction of large systems. Requirements elicitation is not treated in Gaia as in other methodologies.

Regarding urban transportation and its complex system characteristics, we state that there is no AOSE methodology which perfectly models the transport scenario and fulfils its requirements with respect to dynamism and interaction complexity. From the organization's point of view, the available notation is insufficient to represent complex organizations and relations. Thus, a new methodology needs to be developed to take advantage of combined strategies from various areas to overcome the presented issues and this is what our approach aims to achieve.

3.3 Analysis and Design for MAS

Castro and Oliveira proposed, as aforementioned, some new complements to Gaia, and among them, two must be emphasized. First, they use different methodologies, Tropos for requirements elicitation and Gaia to support analysis and design phases. Further, they suggest the use of a service model to clarify all agents' interfaces. However, its service definition is poor because of its origins in the observation of interactions and not in the process flow. We decided to amplify this view and apply Services Oriented Architecture (SOA) concepts instead (Lublinsky, 2007).

SOA can basically define an information system as a group of services that interact to better serve the end user. It aims to create an ad-hoc topology of application that users are allowed to string together as functionality pieces. Therefore, in technical aspects, SOA-based architectures have many advantages. One major advantage is that services are relatively open and accessible by any user or other services, as long as they are able to understand each other. This makes SOA pretty interesting to support services, for instance over the Internet. Yet, within disadvantages, security, and lack of testing and validated models, important issues are still addressable. Secondly, from an economic perspective, the possibilities are great as SOA-based architectures provide the basis for new business opportunities and other similar investments, as well as for the customization of current services making them adaptable to a new emergent dynamic demand (Krämer and Halang, 2007).

Furthermore, we can point out some others values by applying SOA in AOSE. In the service identification SOA uses a combination of top-down, bottom-up, and goal-service modeling (middle-out). The top-down consists of domain decomposition into its functional areas and subsystems. Furthermore, bottom-up observes components in a more reactive way and, finally, middle-out complements the two approaches above interlacing goal and performance indicators (Arsanjani, 2004). So, an AOSE using SOA ensures that, by matching all views, we can cover the issues of complex systems modeling presented earlier.

To illustrate the use of the proposed methodology, we are going to explain it in steps and also instantiate it in a common urban transportation scenario. Scenario description and elicitation of requirements are done; afterwards, we go deeper into the agent, services, and processes definitions. Finally, we end with final diagrams of the detailed design.

3.3.1 Scenario Description and Early Requirements Analysis

Any design process must start by understanding the scenario under study answering some classical questions: What are its components? How do they work and interact with each other? To give this first step we have an elicitation phase composed of two main parts: discovering the scenario work flow and defining an early requirements diagram using Tropos to give a goal-oriented perspective.

To instantiate the methodology, a scenario which is broadly studied by transportation practitioners and the scientific community was chosen. It is a planner system for generating multimodal trips (various types of transportation), started by user request. A simple description is: the user requests the system, through different devices and services, a route

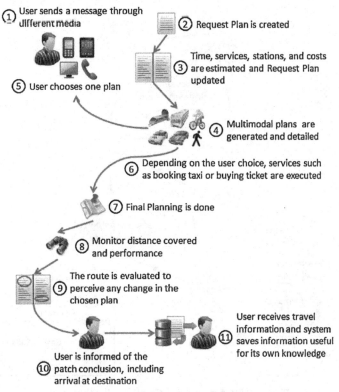

Figure 3.1: Macroflow example for Request Plan

from his/her current point to the desired destination using multimodal transport, and then he/she confirms the best route. Thus, the Request Plan represents the user's desire to have a route to the destination, sent by SMS, email, or phone, to a multimodal planner. The transport scenario (Figure 3.1) is as follows:

1. The Request Plan is created when the user sends a request to the system.
2. For each plan of operation, a second order is created containing all details about time, services, cost, and so forth.
3. All information is updated and checked to ensure that proposed routes meet the desired characteristics.
4. The user chooses the route that most pleases him/her and, depending on it, some tasks are executed such as booking and purchasing tickets.
5. Final planning is made and all subpaths are monitored to generate new routes in case any change in the path is detected.
6. At the conclusion of each subpath information is given to the user, the new checkpoint and, at route completion, some trip information is sent. Also, the system saves key information to improve its own knowledge.

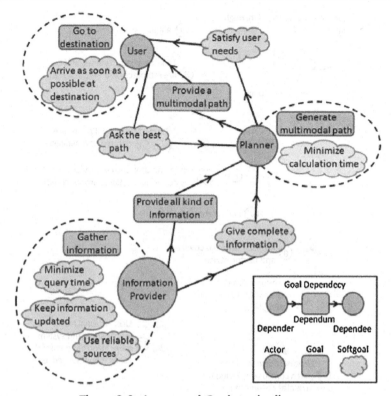

Figure 3.2: Actors and Goals main diagram

An actor/goal-oriented analysis is more efficient, not only to collect requirements, but also to apply the whole methodology, which leads us to a natural view of the scenario similar to what is observed in reality. According to Castro and Oliveira (2008), this approach can help in identifying specific sub-organizations, role basic functionalities, and so forth.

In Figure 3.2 the early requirements diagram using as a basis Tropos methodology is shown. However, before explaining the diagram, we are going to define the used concepts: "actor" is simply an agent, position, or role; "goal" represents what an agent or a group of agents must accomplish during its/their existence; "softgoal"is associated with quality of service, such as safety, security, performance, and so forth" (van Lamsweerde, 2001).

We will begin by explaining the goals and soft goals for each actor, represented by those that are inside the dashed circle. The User has the goal to reach his/her destination, but aims to do this as soon as possible and also he/she does not know whether it will ever occur. Therefore, Planner must generate a multimodal plan that intends to minimize computational time to optimize the overall system response. Also, Information Provider's main objective is to gather information; nevertheless, to improve its service it is complemented with three other soft goals: to use reliable sources, to keep updated information, and to minimize query time.

In addition to individual goals, the model also has shared soft goals and goals to demonstrate that the system must cooperate to accomplish all its goals. Thus, the Information Provider must supply information to the Planner in order to gain performance, which must be the most complete as possible. Furthermore, between User and Planner there exists the goal of providing a multimodal plan and the Planner must still satisfy user needs for, such as explained in the description, the plan must be confirmed by the User. Nevertheless, User request for the best way is a soft goal that does not have necessarily to be fulfilled. The scenario could be more complex but, clearly, the proposal is to instantiate the proposed methodology.

3.3.2 Analysis Phase

Entering at a technical level, we must generate a comprehensive set of analysis models. Our analysis phase, based on Gaia, has five tasks: (1) subdivision of the system into sub-organization, (2) environment model (static and dynamic), (3) preliminary role model, (4) preliminary interaction model, and (5) preliminary service model. A dynamic environment model was added, using Business Process Modeling (BPM) organization model, to improve environment representation. Also, a preliminary service model was introduced encompassing the service concept in the approach. Moreover, some tasks were modified from Gaia into internal levels to complement them at the abstraction level.

As stated in Gaia, we must divide our scenario into suborganizations to better define their goals and make them explicit in the system description. These organizations should exhibit behavior- oriented approach to achieve a given goal, loosely interact with other system's parts, and need similar competences. From requirements elicitation analysis, it is possible to determine three candidate suborganizations that fulfill the above conditions. The suborganizations identified are described in Table 3.1.

Modeling the environment is also one of the agent-oriented methodologies' major concerns. In our case, we divide it in two parts: static that came from Gaia, and dynamic which will

Table 3.1: Identified suborganizations.

Suborganization	Description
User	The subgoal to achieve is "Go to destination". It will only interact with actor Planner because of the dependency "Provide a multimodal path".
Planner	The subgoal to achieve is "Generate multimodal path". It will loosely interact with other portions of the system because of the dependencies "Provide a multimodal path" and "Provide all kind of Information" with the User and Information Provider actors respectively.
Information Provider	The subgoal to achieve is "Gather information". Interaction depends on "Provide all kind of Information" with the Planner actor.

Table 3.2: Resources (Partial).

Name	Description
Users	Contains information regarding all users. This makes it possible to know information about a user, for example, his/her history, frequent destinations, and if he/she is doing a plan now.
Request Plan	Contains information regarding the Request Plan. It allows having information about the origin, destination, possible transport services, and user that made the request.
Transport Timetable	Contains information regarding all Transport available and their timetable. This resource is essential for the system because it contains all timetables and also needs to be frequently updated so it can correspond to the real transport availability.

use a Business Process Modeling Notation (BPMN) (BPMN, 2011) representation. Thus, the first part is called static because it is composed of components that do not change frequently and those are: active (they are entities capable of performing complex operations with which agents can interact); resources (they are data available to compute processes for sensing, effecting, or consuming). Table 3.2 portraits available resources and, in our instantiation, there are no active components.

Complex scenarios, for example, urban transportation, are usually very dynamic. To represent that Gaia uses organization rules. However it is insufficient, so we propose a new approach to represent that using Business Processes (BP). BP tries to model a scenario through a collection of related activities; also we use the BPMN specification by OMG. In Figure 3.3, the collaboration diagram that corresponds to Figure 3.1 scenario is structured.

The actors and goals diagram in Figure 3.2 helps us to identify several roles which will build up the final MAS organization. A partial list of those roles is:

- *PlanChoose*, role associated with deciding what is the best plan to choose, on the basis of user's personal preferences.
- *TransportServicesFind*, role associated with finding all transport services that are possible for generating plans for *PlanChoose* role.
- *InformationGather*, role associated with gathering information from various sources and also different kinds; all these will be used by the *TransportServicesFind* role.

In Table 3.3, we can see the filled preliminary role model for one identified role. Our concept is that roles do not act upon and directly sense the environment; they do it through services making the system modular and flexible. This preliminary role will be complemented and will result in the full role model.

The interaction model aims to capture dependencies and relationships among roles in the MAS organization. It is possible to achieve this by defining protocols for each type of

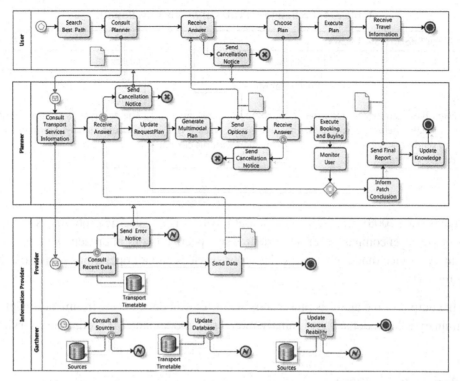

Figure 3.3: Collaboration Diagram for the scenario

inter-role interaction. Table 3.4 shows the definition of a preliminary protocol using the Gaia notation. There are interactions among all roles. However, they will be completely detailed in the design phase.

It is intended to introduce the service concept to agent-oriented methodologies, though the word service is already used in plenty of them (including Gaia) and here however, it follows a different paradigm. Services are generally confused with the operations concept and, to clarify this, service hereafter is defined as a holistic entity that may encapsulate business

Table 3.3: RosterPlanMonitor Preliminary Role.

Role schema: RosterPlanMonitor
Description: This preliminary role involves monitoring the user and some other source for events related to any possible change in the plan (e.g., take different bus). After detecting one of these events the RosterPlanMonitor will request a new plan for PlanGenerate role. It should be able to avoid any abnormality related with requests.
Protocols and activities: CheckForNewEvents, UpdatePlanEventStatus
Responsibilities: Liveness: RosterPlanMonitor = (CheckForNewEvents)^W || (UpdatePlanEventStatus)^W

Table 3.4: Preliminary Protocol *InformTravelEvents*.

Protocol schema: **InformTravelEvents**		
Initiator Role	*Partner Role*	*Input*
PlanApply	RosterPlanMonitor	Position information
Description		*Output*
After an event has been detected it is necessary to find an alternative plan that can make the user reach his/her destination. For that it is necessary to send details about position so a new plan with his current position can be generated.		The new plan is sent taking into account all new information

requirements (Erl, 2008), that is, a set of operations to solve a specific problem. So, in the analysis phase, to encompass a service-oriented perspective in our approach, we must create a preliminary service model presenting the set of services and all operations performed by it (see Figure 3.4).

Finally, the output of the analysis phase consists of four models: (1) environment model, (2) preliminary role model, (3) preliminary interaction model, and (4) preliminary services model.

3.3.3 Architectural Design

The analysis delivers models to express functionality and operations and now, in the design phase, we must complete and refine all preliminary models. Also, it is necessary to make some decisions about a multiagent system that covers all specifications elicited. This intermediary phase aims to complete models observing complementary functions and in the

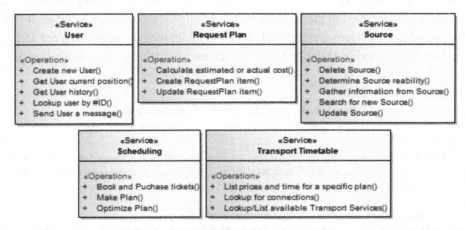

Figure 3.4: Preliminary Service Model

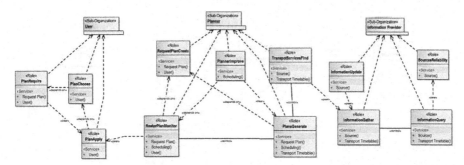

Figure 3.5: Organizational Structure

final diagram it will be generated. We have three tasks: (1) define the organizational structure, (2) complete preliminary models, and (3) make system choreography.

Moreover, from specification analysis we have the organizational structure presented in Figure 3.5. To represent it, Gaia suggests the adoption of a formal notation and a graphical representation. Thus, based on the mapping proposed by Castro and Oliveira, 2008, relating to the Gaia concept in UML 2.0 we generate the diagram shown.

Following the methodology, after defining the organization, the role and interaction preliminary models can be completed. The first task is to insert all protocols in the role model. We added a new field in the Gaia approach to represent the connection among roles and services, which is the required services field. In Table 3.5 the full model for RosterPlanMonitor role is shown.

To complete the interaction model we must observe if, due to the organizational structure, any new protocol is required or even if some modification to the preliminary model is necessary. In our case, the preliminary interaction model does not need to be changed. Nevertheless,

Table 3.5: RosterPlanMonitor Role.

Role schema: RosterPlanMonitor
Description: This preliminary role involves monitoring the user and some other source for events related to any possible change in the plan (e.g., take different bus). After detecting one of these events the RosterPlanMonitor will request a new plan for PlanGenerate role. It should be able to avoid any abnormality related with requests.
Protocols and activities: CheckForNewEvents, UpdatePlanEventStatus, informsNewPlan, requestPlan, requestUserData, reportsTravelQuality.
Required Services: User, Request Plan
Responsibilities: Liveness: RosterPlanMonitor = (CheckForNewEvents)^W || (UpdatePlanEventStatus)^W

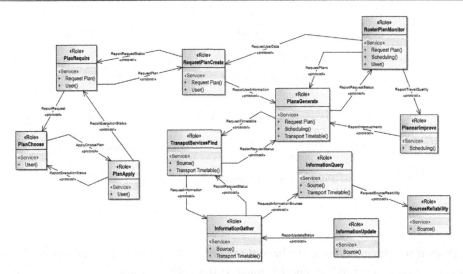

Figure 3.6: Roles and Interactions Diagram

aiming to clarify aspects of roles and interaction, in Figure 3.6, is shown the full model including all protocols and required services that will help us to define which role agents are going to play.

We propose a new modeling approach for services using the aforementioned concept, as represented in Table 3.6. Each service should have its operations described and, correlating it to the static environment model, specify all required resources. For the example instantiated, the operation CreateRequestPlanItem, as it is named, creates the RequestPlan item with different information. Consequently, it has to include an item in the Request Plan data base and connect the item to a User and responsible Roster.

Table 3.6: Request Plan Service Model.

Service schema: Request Plan
Operations
   CalculateEstimatedActualCost: This service calculates the estimated cost for each generated plan and also the actual cost to the current plan to send travel report to the user. It inserts the value in the Plans database to be consulted by other services.
   Resource: Plans Database
   CreateRequestPlanItem: This service creates the RequestPlan item with the user, origin position, destination position, and so forth. It inserts a new item in the Request Plan database after verifying any duplicity and correlates this item to the Users and Roster databases
   Resource: Request Plan, Users, Roster Databases
   UpdateRequestPlanItem: This service updates any new information related to the RequestPlan item, and by doing that also needs to cross information with the Users and Roster databases.
   Resource: Request Plan, Users, Roster Databases

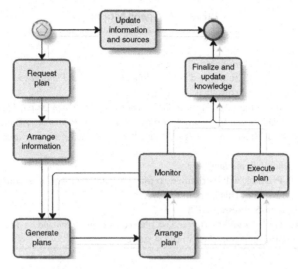

Figure 3.7: Choreography Diagram

The specification of BPMN defines the choreography in two manners: firstly, as a process depicting interactions between business entities, and a sequence of global interaction types of activities; secondly, we use this simpler representation of the collaboration diagram to cross over systems' entities with the ambient dynamics. Figure 3.3 must be generalized and suppress messages exchange so a single process chain can emerge (see Figure 3.7).

The outputs for this methodology phase are: (1) organizational structure, (2) full-role and interaction model, (3) complete services model, and (4) scenario choreography diagram.

3.3.4 Detailed Design

The design phase is the technical kernel to provide the complete system definition. Indeed, the output of this phase is the output of the methodology itself and to fulfill it we must perform two tasks: (1) define the agent model on the basis of the role interaction full model and service the complete model (see Table 3.7), and (2) define the business interaction model to give an overview of the system behavior.

Moreover, even agents requiring services to accomplish their goal must be responsible for making a service available using their own skills in making a flexible structure (cardinality among agents and services can be from many to many). To complete the agent model, we choose to use an UML diagram where agents are presented by classes with the agent stereotype on top of it. This same notation is assumed for services, with the service stereotype instead. The dependency relationship between agents and services with the delegate

Table 3.7: Agent Model.

Agent classes/roles
User[1..n] *play PlanRequire, PlanChoose, PlanApply*
This means that agent class *User* will be defined to play roles *PlanRequire*, *PlanChoose*, and *PlanApply*. It will have between one and n instances of this class in the system.
Planner[1..n] *play RequestPlanCreate, TranspotServicesFind, PlansGenerate, RosterPlanMonitor, PlannerImprove*
This means that agent class *Planner* will be defined to play roles *RequestPlanCreate*, *TranspotServicesFind*, *Plans-Generate*, *RosterPlanMonitor*, and *PlannerImprove*. It will have between one and n instances of this class in the system.
Provider[1..n] *play InformationGather, InformationQuery*
This means that agent class Provider will be defined to play roles *InformationGather* and *InformationQuery*. It will have between one and n instances of this class in the system.
Gatherer[1..n] *play InformationUpdate, SourcesReliability*
This means that agent class *Gatherer* will be defined to play roles *InformationUpdate* and *SourcesReliability*. It will have between one and n instances of this class in the system.

stereotype can be read as "agent x is responsible to perform service y." Figure 3.8 was generated by combining agents, services, and interactions.

Finally, to couple the concepts of agents, services, and processes together and in order to show all interactions among them; we propose the business interaction model (Figure 3.9). This diagram is composed of three main parts: agents (left), BP choreography (centre), and services (right). Thus, it must be read as "an agent requires some service to fulfill a certain process" and all interactions are represented by the red arrows. There is a distinction between

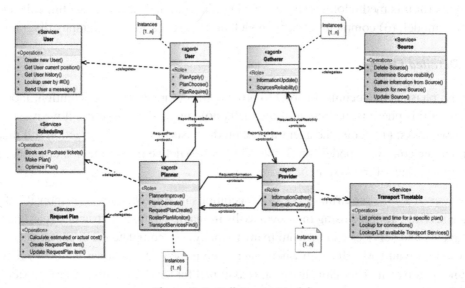

Figure 3.8: Full Agent Model

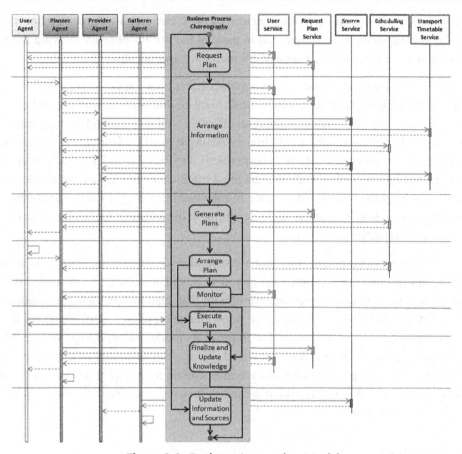

Figure 3.9: Business Interaction Model

arrows: the solid refer to an operation or an action, whereas the dashed ones refer to a message between agents or an acknowledgement to a previous request by an agent.

3.4 Discussion and Future Work

Environmental impacts and safety are, nowadays, two major concerns of the scientific community with respect to transport scenarios and to the ever-growing urban areas. These issues gain more importance due to the increasing amount of vehicles and people. Seeking for new solutions is reaching a point where available technologies and artificial intelligence, especially MAS, are being recognized as ways to cope and tackle these kinds of problems in a distributed and more appropriate way. However, not all solutions can be tested in reality. Indeed, a virtual environment seems to be the best path to follow and for complex systems, such as urban transportation. There are plenty of factors to consider from the very beginning of the modeling phase. Considering that the best way to teach is by examples, we instantiated the proposed methodology in a popular scenario of the transportation domain, namely the multimodal plan generation, and explained in parallel the origins of tasks showing the reason for adding new extensions to traditional MAS modeling approaches.

For this first version, we are focused on two main aspects in our approach. First, the environment model was extended to cover the dynamism that is common in complex systems, for which business processes modeling was an effective solution. Second, we apply the services concept to agent-oriented modeling, expecting it to take advantage of aspects such as modularity (new services can easily be incorporated), flexibility (services can be performed by any agent that has the ability to do it), and allowing the user to play a central role. Due to the various uses of the "service" word, we felt the need to clearly explain the concept and define it more appropriately in our context, resulting in the addition and modification of tasks to fit this novel perspective.

The methodology is not complete and there are several issues remaining open, which need further development and clarification. One is how to model and design in a sound way the data that is a key point in software engineering; with that, also metamodels for databases can be generated to help the developer in the implementation phase. Also, the discussion about service itself is quite extensive and to improve its model we need to go deeper into the service organizational structure and some related issues, such as quality of services, security, trustworthiness will emerge. These aspects are also studied in the agent field as well as in other disciplines, and certainly have the potential to foster a cross-fertilization opportunity in a multidisciplinary fashion.

Acknowledgment

This project has been partially supported by FCT, the Portuguese Agency for R&D (PhD Scholarship grant SFRH/BD/66717/2009).

References

Akbari, O., April 2010. A survey of agent-oriented software engineering paradigm: towards its industrial acceptance. J. Comp. Eng. Res. 1 (2), 14–28.

Arsanjani, A., 2004. Service-oriented modeling and architecture: how to identify, specify, and realize services for your SOA. IBM Syst. J.

BPMN v2.0, January 2011. OMG/Business Process Management Initiative.

Castro, A., Oliveira, E., 2008. The rationale behind the development of an airline operations control centre using Gaia-based methodology. IJAOSE 2 (3), 350–377.

Cernuzzi, L., Juan, T., Sterling, L., Zambonelli, F., 2004. The Gaia methodology – basic concepts and extensions. Methodologies and Software Engineering for Agent Systems: The Agent-Oriented Software Engineering Handbook. Kluwer Academic Publishing, New York, pp. 69-88.

Cossentino, M., Gaud, N., Hilaire, V., Galland, S., Koukam, A., 2010. Aspecs: an agent-oriented software process for engineering complex systems. JAAMAS 20 (2), 260–304, Springer.

Erl, T., 2008. SOA Design Pattern, first ed. Prentice Hall, Boston.

Giunchiglia, F., Mylopoulos, J., Perini, A., 2002. The tropos software development methodology: Processes, models and diagrams. Third International Workshop on Agent-Oriented Software Engineering. pp. 162–173.

Gonzalez-Palacios, J., Luck, M., 2008. Extending gaia with agent design and iterative development. 8th International Workshop on Agent-Oriented Software Engineering (AOSE 2007). Springer. pp. 16–30.

International Transport Forum, 2009. Trends in the Transport Sector 1970-2007. OECD, Paris, June 2009. ISBN 978-92-821-0159-9.

Juan, T., Pearce, A., Sterling, July 2002. Roadmap: Extending the Gaia Methodology for Complex Open Systems. In: Proceedings of the First International Joint Conference on Autonomous Agents and Multiagent Systems (AAMAS '02). ACM, New York, NY, USA. pp. 3–10.

Krämer, B.J., Halang, W.A. (Eds.), 2007. Contributions to Ubiquitous Computing SCI, vol. 42, Springer, Heidelberg.

Kumar, K., Welke, R., 1992. Methodology Engineering R: A Proposal for Situation-Specific Methodology Construction. John Wiley Information Systems, pages 257–269.

Lublinsky, B., 2007. Defining SOA as an architectural style: Align your business model with technology [Online]. < http://www.ibm.com/developerworks/architecture/library/ar-soastyle/>. (Last accessed 18-04-2012)

Odell, J., Parunak, H.V.D., Bauer, B., 2001. Representing agent interaction protocols in uml. First International Workshop on Agent-Oriented Software Engineering (AOSE 2000). Springer-Verlag, New York, Inc. pp. 121–140.

U.S. Environment Protection Agency, 2009. 1970–2008 average annual emissions, all criteria pollutants in ms excel [Online]. <http://www.epa.gov/ttn/chief/trends/trends06/nationaltier1upto2008basedon2005v2.xls>. (accessed June 9, 2009)

van Lamsweerde, A., August 2001. Goal-Oriented Requirements Engineering: A Guided Tour. 5th IEEE international Symposium on Requirements Engineering. Washington. p. 249.

Wang, F-Y., Tang, S., 2005. A framework for artificial transportation systems: from computer simulations to computational experiments. Intelligent Transportation Systems Proceedings. pp. 1130–1134.

Zambonelli, F., Jennings, N.R., Wooldridge, M., July 2003. Developing multiagent systems: the Gaia methodology. ACM Trans. Softw. Eng. Methodol. 12 (3), 317–370.

CHAPTER 4

A Role-Based Method for Analyzing Supply Chain Models

Johan Holmgren*, Linda Ramstedt**, Paul Davidsson†

*Faculty of Computing, Blekinge Institute of Technology, Karlshamn, Sweden, Department of Computer Science, Malmö University, Malmö, Sweden **Sweco, Stockholm, Sweden †Department of Computer Science, Malmö University, Malmö, Sweden

4.1 Introduction

In the global economy, trade takes place in complex networks that stretch over large areas and involve many actors. Managing supply chains is a complex task where many actors interact in the processes of providing products and transport services, and improved supply chain performance is often of high priority for the involved actors. We regard a *supply chain* as the set of activities and organizations that are involved in producing, handling, and moving products from a supplier to a customer. Examples of activities that typically are involved in a supply chain are procurement of raw material, assembling of components, transportation, and terminal activities. A supply chain can be regarded as the complete chain from mining of raw material until delivery of finished products to the end consumer, or as a limited number of steps, for example, in production. In addition, *supply chain management* can be referred to as the planning and management of supply chain activities as mentioned by "Council of Supply Chain Management Professionals" (CSCMP, 2013). In particular, supply chain management is an important means for coordination and collaboration between actors who are involved in the supply chain, such as suppliers, customers, and third-party service providers.

Simulation is for several reasons considered as an important method for improved supply chain management (Tarokh and Golkar, 2006). For example, it can be used to assess the impact of various types of public and private supply chain policy and infrastructure measures. Governmental control policies (e.g., taxes and fees) and infrastructure investments are often used by public authorities to reach governmental goals, such as emission targets and sustainable economic development. Moreover, enterprises often take different types of measures to improve their operations, thereby increasing the profit and strengthening the positions with regard to competitors. For example, the location of a production facility may be changed in order to reduce the need for transportation, or to decrease production costs.

Advances in Artificial Transportation Systems and Simulation.
Copyright © 2015 Zhejiang University Press Co., Ltd. Published by Elsevier Inc. All rights reserved.

To prevent undesirable effects from occurring, it is important to be able to accurately predict the impact of different types of supply chain policy and infrastructure measures. Impact assessment is one important reason for simulating supply chain activities, other purposes are to gain knowledge and understanding of particular complex systems, visualize certain phenomena, and educate persons in the domain.

Various types of techniques have been used for modeling and simulating supply chains. For example, models based on discrete event simulation have been suggested by van der Vorst et al. (2000) and Persson and Olhager (2002), approaches based on system dynamics have been contributed by Minegishi and Thiel (2000), Higuchi and Troutt (2004), and Özbayrak et al. (2007), and approaches based on agent-based modeling have been suggested by Strader et al. (1998), Roorda et al. (2010), and Gjerdrum et al. (2001). Even though supply chain simulation models are based on different modeling techniques and are described in different ways, we argue that it should be possible to describe them in a uniform way that takes particular consideration of their structural characteristics. Since simulation models often are complex, it is believed that a structured way of describing them would facilitate their analysis. Structured analysis is important for different reasons, for example, to determine the scope of a model and to compare different models. Hence, we argue that there is a need for a method that enables structured and uniform analysis of different types of supply chain simulation models.

Supply chains can be organized in different ways, for example, with regard to which organization is responsible for planning and performing different types of supply chain activities. As an example, transportation can in one supply chain be managed by the seller of products, and in other supply chains by the buyer of products, or by a third-party logistics provider. Therefore, it is reasonable to assume that the activities, roles, and responsibilities in different supply chain organizations will be different. However, we argue that on some level of abstraction, it is possible to identify a set of roles, responsibilities, and interactions that are general enough to represent all types of organizations involved in providing and using products and transport services.

We provide a framework of supply chain roles, responsibilities, and interactions, which can be used to describe different types of organizations involved in providing and using products and transport services. We also suggest a method for structured and uniform analysis of supply chain simulation models, which is built upon our framework. In fact, we argue that the method can be useful also when analyzing real world supply chains. Supply chains and their modeling constitute an essential part of an Artificial Transportation System (ATS). Moreover, the modeling is one of the first stages of a simulation project.

It should be emphasized that we limit our focus to describing the organizations that are involved when products and transport services are provided and used, since those are central in most supply chains. That is, our focus is on the modeling of organizations such as suppliers, factories, distribution centers, customers, and providers as well as users of transport services. This means that there exist organizations, such as customs, that are outside of our scope. It should be emphasized here that the concepts of providing and using may refer to something that happens inside an organization, and which do not necessarily have to involve financial transactions. Moreover, they do not have to involve any physical flows, for example, when a third-party logistics provider buys and sells services without using them in their own businesses.

In the next section we discuss some existing approaches to structured analysis of supply chain simulation models. In Section 4.3 we present a framework of supply chain roles, responsibilities, and interactions, which is followed in Section 4.4 by the specification of a method for analyzing supply chain simulation models. In Section 4.5 we give an account of the applicability and validity of the framework and method, and in Section 4.6 we provide some concluding remarks and directions for future work.

4.2 Related Work

When analyzing and comparing simulation models a common approach is to make use of some kind of categorization framework or data extraction form to make sure that models are analyzed in a uniform way. Below, we briefly describe a couple of typical categorization frameworks that have been used for structured analysis of simulation models. Even though they do not focus on identifying the structure of models, we discuss them here since they cover relevant areas of application and they focus on uniform analysis of simulation models.

Terzi and Cavalieri (2004) provided a survey of supply chain simulation models, in which they reviewed around 80 articles by making use of a structured categorization framework. The framework covered several aspects related to scope and objectives, simulation paradigm and technology, and development stage of simulation models. Chow et al. (2010) analyzed the state-of-the-art of simulation-based approaches to freight forecast modeling by making use of an analysis framework focusing on how different types of forecasting models are built and how they are able to fulfill different types of needs for analysis. In particular, the framework covered which components are included in an analyzed model, for example, trip generation, trip distribution, and traffic assignment, and the types of the model, for example, economic flow factor model, commodity model and economic based model.

In another survey Davidsson et al. (2007) presented a framework that was used for structured analysis of agent-based simulation models. Even though the focus in the survey by Davidsson et al. was on a rather wide range of application areas it is relevant for our work since its purpose was to analyze agent-based models. The analysis framework covered several aspects related to problem description, modeling, implementation, and results. In particular, the categorization framework contained a category for identifying the structure of an agent system by categorizing the architecture as peer-to-peer, hierarchical, or centralized. Examples of questions that can be answered by using the framework by Davidsson et al. for analyzing an agent-based simulation model are:

- Who is the intended end user of the model?
- What is the purpose of the model?
- What is the maturity of the model?
- What is the typical scale of a simulation study with the analyzed model?
- How has the model been validated?

Even though approaches exist for structured analysis and categorization of supply chain simulation models, we have observed that most of them focus on extraction of data as well as for categorization of models. We have not found any approach that is based on agent-based concepts such as roles, responsibilities and interactions for analyzing the structure of models. We believe that an analysis method based on these concepts may provide improved understanding of the structure of supply chain simulation models.

The presented approach for analyzing supply chain simulation models is inspired by Gaia (Zambonelli et al., 2003), which is an agent-based methodology for analysis and design of agent systems. It could be possible to build a structured analysis method around other general-purpose agent-based development methodologies, such as Tropos (Bresciani et al., 2004) and Promotheus (Padgham and Winikoff, 2004). It could also be possible to build an analysis method around the Multi-Agent Supply Chain Framework (MASCF) (Govindu and Chinnam, 2007), which is a methodology that can be used in the analysis and design phases for facilitating the development of agent-based supply chain models. MASCF is based on the Gaia methodology, to which it applies the Supply Chain Operations Reference model (SCOR), which is a framework of standard supply chain processes, metrics, best practices, and so forth.

Next, we briefly describe a number of model-based frameworks for developing agent-based supply chain models. Swaminathan et al. (1998) presented a modeling framework that defines two types of supply chain elements: (1) structural elements (e.g., retailers, manufacturers, and transporters) are modeled as agents that communicate with each other via specific interaction protocols, and (2) control elements (e.g., inventory policies) are used by agents to coordinate the flow of products. Chatfield et al. (2007) suggested a conceptual architecture for construction of supply chain simulation models. The architecture follows an order-centric view and it contains a knowledge base, a simulation framework, a conversion process, and a simulation environment. The knowledge base defines five types of constructs (also referred to as fundamental classes), that is, nodes, arcs, components, actions, and policies. In the architecture, reactive agents are used to represent nodes and arcs. Another simulation modeling framework was proposed by van der Zee and van der Vorst (2005). Their framework was built upon three basic classes: (1) agents represent supply chain entities (e.g., planners, production departments and distribution systems), (2) flow items define movable items in a supply chain (e.g., products and resources), and (3) jobs are used to link together agents and flow items.

Even though these approaches are targeted toward designing models, while we are focusing on structured analysis, we found it relevant here to briefly describe them. The main reason is that they describe systems by making use of structured agent-based architectures and frameworks. In addition to designing models, it could be possible to use them also for analyzing supply chain simulation models. However, we suggest a completely different approach that it is based on abstract roles, responsibilities, and interactions. In particular, our approach does not make use of traditional organizational roles, such as producers, transport operators, customers, and different types of planner roles that are defined internally in organizations. Since an organization may represent multiple roles in a supply chain, for example, producer and customer, instead we define roles as something that describe how different supply chain organizations relate to each other, that is, as providers and users of products and transport services. We believe this is an appropriate approach to facilitate the understanding of the structure of supply chain simulation models.

4.3 A Framework of Supply Chain Roles, Responsibilities and Interactions

We will now describe a framework based on supply chain roles, responsibilities, and interactions, that is, concepts commonly used to describe agents. Agents are often used for modeling of intelligent and autonomous systems, and can be described as computer systems that are capable of autonomous, reactive, goal-directed, and social behavior in their environment of operation, in order to meet their design objectives (Wooldridge and Jennings, 1995).

Before introducing the framework, a few important concepts will be introduced. As mentioned in the introduction, a supply chain can be defined as the activities and organizations that are involved in producing, handling, and moving products from a supplier to a customer. In a supply chain, organizations refer to customers, producers, transport operators, and so forth. Organizations may also refer to suborganizations within larger organizations, which enable internal representation of supply chains, for example, when production is performed in multiple steps. We regard an organization as a set of roles, such as transport planner and order administrator. A role is often defined as a set of responsibilities, and in the context of decision-making, a responsibility can be viewed as a set of decisions. Different organizations typically include different roles, and just as roles are assigned to humans in the real world, roles are in an agent system assigned to agents.

To illustrate how we make use of the concepts of organizations, roles, responsibilities and decisions, we describe a supply chain (SC) as a set of organizations.

$$SC = \{or\ g_1, \ldots or\ g_1\}$$

An organization *org* contains a set of roles

$$Rol_{org} = \{rol_1, \ldots, rol_j\},$$

and a set of responsibilities

$$Resp_{org} = \{resp_1, \ldots, resp_k\},$$

where a responsibility $resp_k$ is defined as a set of decisions

$$resp_k = \{dec_{k1}, \ldots dec_{kL}\}.$$

4.3.1 Roles

In our framework, roles are considered as something that links together organizations. Due to the focus of our work, organizations may relate to each other either as providers or users of products and transport services. Hence, there are four supply chain roles, which we argue can be used to represent all types of organizations involved in providing and using products and transport services:

- Transport User (*TU*). A consumer, or a buyer, of transport services, that is, someone who wants to have products transported.
- Transport Provider (*TP*). A provider or seller of transport services.
- Product User (*PU*). A buyer or user of products.
- Product Provider (*PP*). A provider, or a seller, of products.

We let the subset $Rol_o \subseteq \{TU_o, TP_o, PU_o, PP_o\}$ denote the representation of the four roles in an organization *o*. It should be emphasized here that it is not always the case that all four roles are represented in an organization. For example, a producer maybe only plays the role of a *PP*. However, an organization is allowed to, and typically will, play more than one role. For example, a freight forwarder plays the roles of *TU* and *TP*. It provides transport services to *TU*:s and it consumes transport services from *TP*:s.

4.3.2 Responsibilities

As mentioned above, a role is often defined as a set of responsibilities, and we have chosen to regard a responsibility as a set of decisions related to some specific aspect or activity. To capture the processes of ordering products and transport services, we have identified two main types of decisions that, in different ways, need to be represented in supply chain organizations involved in providing and using products and transport services. *Ordering decisions* concern ordering of products and transport services, and include choice of provider, quantities to order, and so forth. *Planning decisions* concern planning of resources related to production, transportation, and inventories. Ordering decisions are typically based on ordering policies, such as the Economic Order Quantity (EOQ) model (Wilson, 1934), which takes into account costs for ordering and transportation. Planning decisions take into account the availability of resources (production facilities, vehicles, etc.) and are typically based on different strategies, for example, concerning scheduling and load coordination.

It should be noted that even though a responsibility is defined as a set of decisions, there are other more concrete activities, such as the actual planning processes, that need to be included in a responsibility to enable good decisions to be taken. However, in our framework we only consider decisions since other activities are implicitly captured when identifying the decisions. Moreover, it should be mentioned that we focus on the planning level; the operational level is outside the scope of our framework. We, therefore, refer to all types of responsibilities as planning responsibilities.

We have identified four planning responsibilities that can be used to represent the decision-making in different types of organizations involved in providing and using products and transport services. These four responsibilities are presented below together with the main decisions (formulated as questions) that are included in each type of responsibility.

- Product order planning (*POP*). What types of products should be ordered, and in what quantities? When should products be ordered? In what time window should products be delivered? From which product provider should products be ordered? What is the appropriate level of inventory for each product type (taking into account the costs for ordering and inventory holding)?
- Transport order planning (*TOP*). From which transport provider(s) should transports be ordered? What quantity should be transported, and in what time windows should pickup and delivery occur?
- Production resource planning (*PRP*). In which production facility should products be produced? When should products be produced, and in what quantities? How should production be scheduled? To fulfill an order, how many products should be taken from inventory, and how many should be produced? Which inventories should be replenished by production, and in what quantities?
- Transport resource planning (*TRP*). Which transport modes, vehicle types, load carriers and routes should be used? What consignment size is appropriate and when should transportation take place? Which products should be load-coordinated? Should transportation be performed according to timetables or not?

Which responsibilities are represented in an organization depends on which roles the organization plays in the supply chain. We have identified a mapping between the four responsibilities and the four roles: product order planning maps to the product user (i.e., $POP \rightarrow PU$), production resource planning maps to the product provider (i.e., $PRP \rightarrow PP$), transport order planning maps to the transport user (i.e., $TOP \rightarrow TU$), and transport resource planning maps to the transport provider (i.e., $TRP \rightarrow TP$). For future reference we let the subset $Resp_o \subseteq \{POP_o, PRP_o, TOP_o, TRP_o\}$ denote the responsibilities that are represented in an organization (type) $o \in Org$. It has to be assumed that responsibilities are disjoint, that is, that the same decision is assigned only to one responsibility.

Moreover, which particular decisions that define a responsibility depend on the characteristics of the particular organization. The same type of responsibility may involve different decisions when it is represented in different organizations. It may also be the case that the same type of decision will belong to different responsibilities in different organizations. An example is inventory planning, which may be captured both in the product order planning responsibility and in the production resource planning responsibility.

In the framework we also capture the possibility to delegate decisions, which means that responsibilities completely or partially are transferred from one organization to another. A typical example is Vendor Managed Inventory (VMI) (Daugherty et al., 1999), in which decisions about replenishments of customer inventories are delegated to the supplier. In the terminology used here, this means that the product order planning responsibility is delegated from the product user (in the customer) to the supplier.

4.3.3 Interaction

We have identified two types of interaction that are included in our framework, that is, interaction between roles and interaction between responsibilities. Since the framework is defined on an abstract level, the identification of interaction is limited to identifying connections (interaction links) between the roles and responsibilities that need to interact.

The first type of interaction is referred to as *role interaction* and it is used to represent ordering interaction between organizations. The interaction that occurs between organizations can be captured by identifying connections between

1. product provider and product user roles in those organizations that may exchange products, and
2. transport provider and transport user roles in those organizations that may exchange transport services.

The second type of interaction is referred to as *responsibility interaction* and it is used for representing coordination that enables decisions, or responsibilities, to be coordinated. Coordination in decision-making can be recognized as something that occurs between responsibilities in order to allow related decisions to be taken. An example of what is considered to be coordination is coordination of production resource planning and transport resource planning, for example, in order to reduce the need for inventory buffers in a supply chain. Responsibility coordination mainly occurs inside organizations, but it can also represent coordination of decisions that belongs to different organizations.

The responsibility (coordination) interaction that occurs in an organization can be described as a subset

$$Int_{o'}^{resp} \subseteq \left\{ \{resp_i, resp_j\} : resp_i, resp_j \in Resp_{o'}, resp_i \neq resp_j \right\}$$

of the set of all unordered pairs of responsibilities in o'. In the same way, the responsibility (coordination) interaction that occurs between two organizations o' and o'' can be described as a subset

$$Int_{o',o''}^{resp} \subseteq \left\{ \{resp_i, resp_j\} : resp_i \in Resp_{o'}, resp_j \in Resp_{o''} \right\}.$$

The role interaction that occurs between two organizations o' and o'' can be described as a subset

$$Int_{o',o''}^{role} \subseteq \left\{ \{PP_{o'}, PU_{o''}\}, \{PU_{o'}, PP_{o''}\}, \{TU_{o'}, TP_{o''}\} \right\}.$$

4.3.4 An Illustrating Example

After introducing all components in our framework, that is, roles, responsibilities, and interactions, in Figure 4.1 we provide an example of how the framework can be applied. We illustrate how a system containing a customer and a supplier of products is represented in our framework under two different ordering and replenishment policies, that is, normal ordering and VMI (vendor managed inventory). Even though the example is rather small, it captures the most important aspects in our framework. It includes all types of roles, responsibilities, and interactions that were defined in the framework, and it also captures delegation of responsibilities.

In the normal ordering process, the customer sends orders to the supplier based on some ordering policy, and the supplier is responsible for planning production and ordering transports for delivery of products. In the VMI case, the customer only provides information (e.g., consumption forecasts and inventory levels) to the supplier. The product order planning responsibility is in the VMI case delegated to the supplier. In both cases the transport operator organization plans its transport resources and provides transport services to the supplier. Under both policies, the responsibilities of the supplier need to be coordinated to achieve good overall system performance.

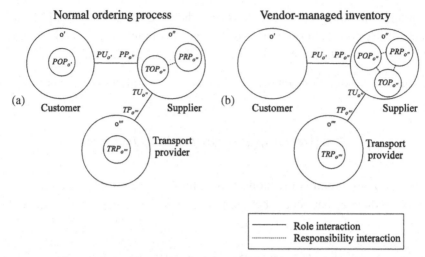

Figure 4.1: Illustration of how a system containing a customer and a supplier of products is represented in our framework under two different ordering and replenishment policies: (a) a "normal" ordering process and (b) VMI. The system also contains a transport provider organization that provides transport services. Under both policies, the supplier is assumed to be responsible for ordering transports, and transport order planning needs to be coordinated with the other responsibilities of the supplier. In the VMI situation, the supplier is responsible for planning its own inventory, as well as the inventory of the customer, since the product order planning responsibility has been delegated to the supplier.

For the two ordering and replenishment strategies that are illustrated in Figure 4.1, we get the following notation:

a. Normal ordering process.
   - Roles. $Rol_{o'} = \{PU_{o'}\}, Rol_{o''} = \{PP_{o''}, TU_{o''}\}, Rol_{o'''} = \{TP_{o'''}\}.$
   - Responsibilities. $Resp_{o'} = \{POP_{o'}\}, Resp_{o''} = \{PRP_{o''}, TOP_{o''}\},$
   - Responsibility interaction. $Int_{o'}^{resp} = \varnothing, Int_{o''}^{resp} = \{\{PRP_{o''}, TOP_{o''}\}\}, Int_{o'''}^{resp} = \varnothing.$
   - Role interaction. $Int_{o',o''}^{role} = \{\{PP_{o''}, PU_{o'}\}\}, Int_{o'',o'''}^{role} = \{\{TP_{o'''}, TU_{o''}\}\}.$

b. Vendor managed inventory
   - Roles. $Rol_{o'} = \{PU_{o'}\}, Rol_{o''} = \{PP_{o''}, TU_{o''}\}, Rol_{o'''} = \{TP_{o'''}\}.$
   - Responsibilities. $Resp_{o'} = \varnothing, Resp_{o''} = \{POP_{o''}, PRP_{o''}, TOP_{o''}\}, Resp_{o'''} = \{TRP_{o'''}\}.$
   - Responsibility interaction. $Int_{o'}^{resp} = \varnothing,$

$$Int_{o''}^{resp} = \{\{POP_{o''}\}, \{POP_{o''}, TOP_{o''}\}, \{PRP_{o''}, TOP_{o''}\}\}, Int_{o'''}^{resp} = \varnothing.$$

   - Role interaction. $Int_{o',o''}^{role} = \{\{PP_{o''}, PU_{o'}\}\}, Int_{o'',o'''}^{role} = \{\{TP_{o'''}, TU_{o''}\}\}.$

4.4 A Method for Analyzing Supply Chain Simulation Models

We here present a role-based method for analyzing various types of supply chain simulation models. The method is defined as five activities that act as guidance when systematically applying our framework of supply chain roles, responsibilities, and interactions to a supply chain simulation model. To facilitate the analysis processes and for communicating results we suggest using graphical illustrations (see Section 4.5 for illustrative examples) whenever possible. Further, we suggest that the activities should be accomplished in the sequence they are specified below.

Activity 1. Identify all supply chain organizations and suborganizations (or types of organizations in case there are multiple identical organizations) *Org* that are involved in providing and using products and transport services.

Activity 2. For each organization (type) $o \in Org$, identify which of the four roles it represents, that is, create the subset $Rol_o \subseteq \{TU_o, TP_o, PU_o, PP_o\}$.

Activity 3. For each organization $o \in Org$, identify which decisions are modeled in o and map them to the different planning responsibilities, that is, create the subset $Resp_o \subseteq \{POP_o, PRP_o, TOP_o, TRP_o\}$.

Activity 4. Identify which organizations (or organization types) that may exchange products and transport services, that is, define which roles are connected. This guideline concerns creating the set $Int_{o',o''}^{role}$ for each pair of organizations $o', o'' \in Org$.

Activity 5. Identify responsibilities that need to be coordinated (i.e., responsibilities with decisions that need to be coordinated), that is, for each organization $o' \in Org$ define the set $Int_{o'}^{resp}$ and for each pair of organizations $o', o'' \in Org$, create the set $Int_{o',o''}^{role}$.

Our analysis method is inspired by the earlier phases of Gaia (Zambonelli et al., 2003), which is a methodology for analysis and design of agent systems that makes use of the same agent-based concepts as our framework is based upon. Gaia identifies roles, responsibilities, and interactions in a similar way as suggested by our analysis method. We also share with Gaia the idea about taking an organizational view of the system to be analyzed.

4.5 Applicability and Validity of the Framework and Analysis Method

We here give an account of the applicability and validity of our analysis method by describing how we have used it to analyze five different supply chain simulation models. In Section 5.1 we describe how we analyzed TAPAS (Holmgren et al., 2012), which is an agent-based simulation model for quantitative impact assessment of transport policy and infrastructure measures. In Section 5.2 we present how we analyzed an agent-based simulation model by Strader et al. (1998), which has been used to study the impact of information sharing in a supply chain network. After that, in Section 5.3 we describe our analysis of a model by Gjerdrum et al. (2001), which is based on optimization and agent-based modeling. In Section 5.4 we describe how we analyzed a model for discrete event simulation of a chilled food supply chain, which was contributed by van der Vorst et al. (2000). Finally, in Section 5.5 we present an analysis of a system dynamics model by Higuchi and Troutt (2004), for simulation and analysis of a short life cycle product supply chain.

4.5.1 Analysis of the TAPAS Simulation Model

We first describe our analysis of an agent-based simulation model called TAPAS (Transportation and Production Agent-Based Simulator) (Holmgren et al., 2012). TAPAS is a micro-level simulation model for quantitative impact assessment of transport chain policy and infrastructure measures, and it works by simulating how different types of supply chain actors are expected to act under different conditions. The following six supply chain roles are modeled as agent types: (1) customer (C), (2) supply chain coordinator (SCC), (3) product buyer (PB), (4) production planner (P), (5) transport buyer (TB), and (6) transport planner (T).

The agents are assumed to follow a predefined interaction protocol for matching production and transportation that fulfill customer orders. The progress of TAPAS is driven by consumption, and the interaction protocol is initiated when a C sends an order request with a set of different order quantities to the SCC. After that, the SCC asks the PB to ask the P:s for relevant product proposals, and for each of the received proposals it asks the TB for matching transport solutions for a set of known routes. The TB generates transport solutions by asking the T:s for transport proposals for relevant parts of the routes, and the transport proposals are then combined into overall transport solutions. For each of the requested quantities, the SCC chooses the least cost combination of products and transportation, and informs the C about its choice. Then the C chooses the quantity, which from a cost perspective is the most beneficial, when minimizing the costs for storage and ordering. Finally, the chosen alternative is booked with all the involved P:s and T:s. Hence, the agents are organized hierarchically as shown in Figure 4.2.

Activity 1. It can be argued that the six roles that in TAPAS are modeled as agents can be found either implicitly or explicitly in all supply chains. More than one role is typically represented in the same organization, and since TAPAS enables modeling of different types of supply chains we let in our analysis each agent type correspond to a supply chain organization type, that is, C, SCC, PB, P, TB and T.

Activity 2. Our analysis suggested that the following roles are represented in the six identified types of organizations:

- Customer. $Rol_C = \{PU_C\}$.
- Product buyer. $Rol_{PB} = \{PP_{PB}, PU_{PB}\}$.

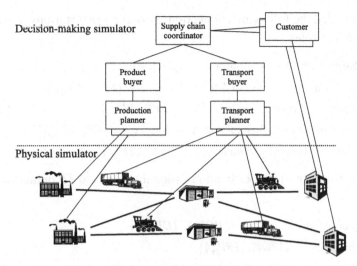

Figure 4.2: Illustration of the architecture of the TAPAS simulation model, with a decision-making simulator modeled by agents and a physical simulator, which is based on an object-oriented paradigm.

- Production planner. $Rol_P = \{PP_P\}$.
- Supply chain coordinator. $Rol_{SCC} = \{PP_{SCC}, PU_{SCC}, TU_{SCC}\}$.
- Transport buyer. $Rol_{TB} = \{TP_{TB}, TU_{TB}\}$.
- Transport planner. $Rol_T = \{TP_T\}$.

Activity 3. The main responsibility for a customer is product-order planning, which implicitly also involves inventory planning. The *SCC* is responsible for product order planning, for transport order planning, and for coordination of production and transportation. The *P*:s perform production resource planning, which also involves inventory planning, and the *T*:s are involved in transport resource planning. The *TB* deals with transport order planning and with transport resource planning, however not the same type of transport resource planning as the *T*:s are dealing with. The *T*:s plan the vehicles, while the *TB* combines transports into overall transport solutions. The responsibilities that are modeled in the identified organization types are summarized as:

- Customer. $Resp_C = \{POP_C\}$.
- Product buyer. $Resp_{PB} = \{POP_{PB}\}$.
- Production planner. $Resp_P = \{PRP_P\}$.
- Supply chain coordinator. $Resp_{SCC} = \{POP_{SCC}, TOP_{SCC}\}$.
- Transport buyer. $Resp_{TB} = \{POP_{TB}, TRP_{TB}\}$.
- Transport planner. $Resp_T = \{TRP_T\}$.

Activity 4. We identified the following connections between organizations, which are represented by pairs of roles:

- Customer – Supply chain coordinator. $Int^{role}_{C,SCC} = \{\{PP_{SCC}, PU_C\}\}$.
- Product buyer – Supply chain coordinator. $Int^{role}_{PB,SCC} = \{\{PP_{PB}, PU_{SCC}\}\}$.
- Production planner – Product buyer. $Int^{role}_{P,PB} = \{\{PP_P, PU_{PB}\}\}$.
- Supply chain coordinator – Transport buyer. $Int^{role}_{SCC,TB} = \{\{TP_{TB}, TU_{SCC}\}\}$.
- Transport buyer – Transport planner. $Int^{role}_{TB,T} = \{\{TP_T, TU_{TB}\}\}$.

Activity 5. We identified internal coordination inside the *SCC* and *TB* organization types, that is:

- Supply chain coordinator. $Int^{resp}_{SCC} = \{\{POP_{SCC}, TOP_{SCC}\}\}$.
- Transport buyer. $Int^{resp}_{TB} = \{\{TOP_{TB}, TRP_{TB}\}\}$.

A graphical illustration of how the TAPAS simulation model is represented in our framework is given in Figure 4.3.

4.5.2 Analysis of a Supply Chain Model by Strader et al. (1998)

Next we describe our analysis of an agent-based simulation model by Strader et al. (1998), which has been used to show the importance of information technology for supporting the order fulfillment process, by studying a 5-tier supply chain network under different demand management and information sharing policies.

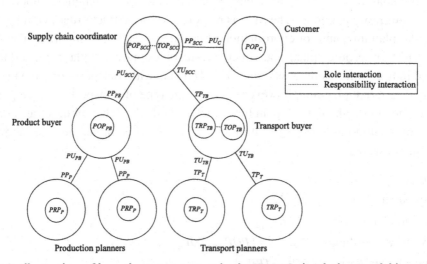

Figure 4.3: Illustration of how the agent system in the TAPAS simulation model is represented in our framework.

Two types of entities are represented in the model, that is, (1) entities with manufacturing (assembling) capabilities and (2) entities without manufacturing capabilities, hence called distribution centers. The two entity types can be combined in different ways to represent different types of supply chains and supply chain networks. The entities are represented by eight agent types: (1) order management, (2) supply chain network management, (3) inventory management, (4) material planning, (5) production planning, (6) shopfloor control, (7) manufacturing systems, and (8) capacity planning agents. The main difference between entities with and without manufacturing capabilities is that, in an entity without manufacturing capabilities only order management, supply chain network management, and inventory management agents are represented, whereas in a manufacturing entity, also material planning, production planning, shopfloor control, manufacturing systems, and capacity planning agents are represented.

Activity 1. We let the two types of supply chain entities that are included in the model be represented by two organization types; manufacturing entity (M) and distribution center (DC), where (DC):s represent entities without manufacturing capabilities.

Activity 2. Depending on how the two types of entities (organization types) are represented in a supply chain network, both of them may act as providers and buyers of products. This can be summarized as:

- Manufacturing entity. $Rol_M = \{PP_M, PU_M\}$.
- Distribution center. $Rol_{DC} = \{PP_{DC}, PU_{DC}\}$.

Activity 3. Our analysis suggested that the two organizations types contain the following responsibilities:

- Distribution center. $Resp_{DC} = \{POP_{DC}\}$.
- Manufacturing entity. $Resp_M = \{POP_M, PRP_M\}$.

Activity 4. Depending on how a particular supply chain or supply chain network is defined, both types of supply chain organizations can represent providers as well as buyers of products. The possible external role interaction links that were identified are denoted:

- $Int_{DC,M}^{role} = \{\{PP_{DC}, PU_M\}\}, \{PP_M, PU_{DC}\}$.

Activity 5. We only identified coordination within the manufacturing entity organization type, that is:

- Manufacturing entity. $Int_M^{resp} = \{\{POP_M, PRP_M\}\}$.

In Figure 4.4, we illustrate how the model is represented in our framework. It should be noted that the model does not explicitly capture transportation between supply chain entities, which is why we chose not to consider transportation in our analysis.

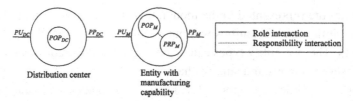

Figure 4.4: Illustration of how the model by Strader et al. (1998) is represented in our framework.

4.5.3 Analysis of a Supply Chain Model by Gjerdrum et al. (2001)

Next we describe how we analyzed an approach to supply chain modeling and performance analysis that was proposed by Gjerdrum et al. (2001). The approach is based on optimization and agent-based modeling, and it defines a hierarchical agent system, which is illustrated in Figure 4.5. The agent types included in the model can be described in the following way:

- Customer agent. It sends order requests to the warehouses, and after receiving cost quotes (from the warehouses), it chooses the most beneficial warehouse for fulfilling the requested order.
- External logistics agent. It acts as an intermediary between the customer and the warehouses.

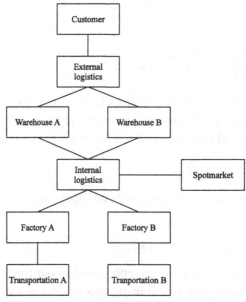

Figure 4.5: Architectural overview of the agent system in the model by Gjerdrum et al. (2001).

- Warehouse agent. When a warehouse agent receives an order it either sends a confirmation to the customer or it requests an emergency order from the spot market depending on the current inventory level. The warehouse agent regularly checks its inventory levels against the order points and, if needed, it orders more products from the factories and from the spot market.
- Internal logistics agent. Operates as an intermediary between the warehouses, the factories, and the spot market.
- Spot market agent. Receives order requests and checks whether or not it can fulfill the requests for products.
- Factory agent. Just as for the spot market, a factory agent can fulfill requests for products. The factory agents are equipped with optimization-based resource planning software to enable optimization of their production resources, and it sends order requests to its transportation agent.
- Transportation agent. Each factory has one transportation agent who is responsible for deliveries.

Activity 1. From the description provided by Gjerdrum et al. (2001) it is not obvious how the modeled agents form different supply chain organizations. In particular, it is not clear if the external logistics agent belongs to the same organization as the customer or if it represents a separate organization. Therefore, in our analysis, it was reasonable to let each agent type define a separate organization type. Hence, we defined the following seven organization types: customer (C), external logistics (EL), warehouse (W), internal logistics (IL), spot market (S), factory (F), and transportation (T).

Activity 2. Our analysis suggested that the identified organization types represent supply chain roles in the following way:

- Customer. $Rol_C = \{PU_C\}$.
- External logistics. $Rol_{EL} = \{PP_{EL}, PU_{EL}\}$.
- Factory. $Rol_F = \{PP_F, TU_E\}$.
- Internal logistics. $Rol_{IL} = \{PP_{IL}, PU_{IL}\}$.
- Spot market. $Rol_S = \{PP_S\}$.
- Transportation. $Rol_T = \{TP_T,\}$.
- Warehouse. $Rol_W = \{PP_W, PU_W\}$.

Activity 3. We identified that the following responsibilities are represented in the model:

- Customer. $Rol_C = \{POP_C,\}$.
- Factory. $Resp_F = \{POP_F, TOP_F\}$.
- Internal logistics. $Resp_{IL} = \{POP_{IL},\}$.
- Warehouse. $Resp_W = \{POP_W\}$.

The external logistics organization (agent) is described as a distributor of messages between the customer and the warehouses. Therefore, we chose not to consider any responsibilities in the external logistics organization. In addition, in the spot market and transportation organization types, no internal operations are explicitly represented in the model, and we chose not to include any responsibilities in these organization types.

Activity 4. In our analysis, we identified the following connections between roles in different organizations:

- Customer – External logistics. $Int_{C,EL}^{role} = \{\{PP_{EL}, PU_C\}\}$.
- External logistics – Warehouse. $Int_{EL,W}^{role} = \{\{PP_W, PU_{EL}\}\}$.
- Factory – Transportation. $Int_{F,T}^{role} = \{\{TP_T, PU_F\}\}$.
- Internal logistics – Factory. $Int_{IL,F}^{role} = \{\{PP_F, PU_{IL}\}\}$.
- Internal logistics – Spot market. $Int_{IL,S}^{role} = \{\{PP_S, PU_{IL}\}\}$.
- Internal logistics – Warehouse. $Int_{IL,W}^{role} = \{\{PP_{IL}, PU_W\}\}$.

Activity 5. We only identified coordination within the factory organization type, that is:

- Factory. $Int_F^{resp} = \{\{POP_F, PRP_F\}, \{POP_F, TOP_F\}, \{PRP_F, TOP_F\}\}$.

Due to absence of details in the description of the model by Gjerdrum et al. (2001) we had to make assumptions regarding how responsibilities are coordinated. Nevertheless, we argue that our analysis, which is graphically illustrated in Figure 4.6, provides a good insight into the structure of the model.

From the analysis we noticed, for example, that no customer inventories are modeled. This seems reasonable as the modeled supply chain can be assumed to start when customer orders are communicated to the external logistics agent. What is more remarkable is that transportation and transport planning is considered only by stating that the transportation agent belonging to a factory is responsible for transferring products to the warehouses. Who is responsible for transporting products from the spot market and from the warehouses is not discussed. It is also not discussed who plans for these transports. A possibility for letting transportation have a more central role in the approach would be to let all transports be provided by one or more third-party transport operators, who would be responsible for planning, coordinating, and providing all transports in the supply chain network. Assuming that transportation always is ordered by the provider of products, the effect this would have on the model is that warehouse agents, factory agents and the spot market agents would be assigned the role *TU* and responsibility *TOP*, while the transport operator agents would be assigned the role *TP* and responsibility *TRP*.

4.5.4 Discrete Event Simulation of a Food Supply Chain

We here describe our analysis of a simulation model by van der Vorst et al. (2000), which has been used to analyze different designs for a chilled salads supply chain. The approach,

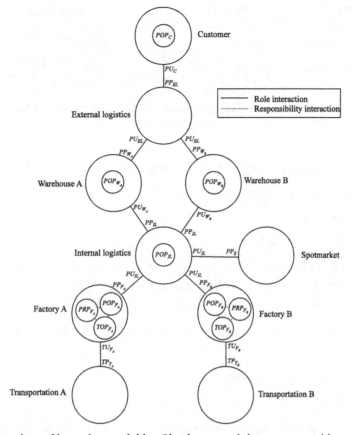

Figure 4.6: Overview of how the model by Gjerdrum et al. is represented in our framework.

which is based on discrete event simulation and petri-net modeling, takes a process-oriented viewpoint and the main components are business processes, design variables, and business entities. A business process is viewed as a set of activities that should be accomplished in a specific order and according to some specific policy, design variables are used to manage and control business processes, and a business entity is something that flows in the supply chain. As an example, production is a business process, design variables determine the production policy, and the actual products are business entities. An overview of the business processes included in the model is given in Figure 4.7.

Activity 1. We identified three (types of) organizations that are represented in the model, that is, (1) one producer (*P*), (2) one distribution center (*DC*), and (3) two retail outlets (*RO*). In the model, the retail outlets are specified in such a way that they represent 100 retail outlets in the modeled real world supply chain.

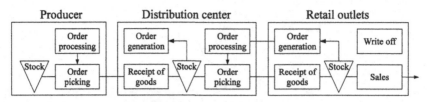

Figure 4.7: Overview of the business processes included in the model by van der Vorst et al. (2000).

Activity 2. Our analysis suggested that the following roles are represented in the model:

- Distribution center. $Rol_{DC} = \{PP_{DC}, PU_{DC}\}$.
- Producer. $Rol_{P} = \{PP_{P}\}$.
- Retail outlet. $Rol_{RO} = \{PU_{RO}\}$.

It should be noted that the retail outlets are regarded as end points in the supply chain. A retail outlet provides products to the customers (the end-consumers). However, it is modeled in a way that we in the analysis chose not to represent it as a product provider.

Activity 3. Our analysis suggested that the following responsibilities are represented in the model:

- Distribution center. $Resp_{DC} = \{POP_{DC}\}$.
- Retail outlet. $Resp_{RO} = \{POP_{RO}\}$.

The model does not explicitly capture planning of inventory and production by the producer, and therefore, we did not include these aspects in the analysis.

Activity 4. We identified the following connections between roles:

- Distribution center – Producer. $Int_{DC,P}^{role} = \{\{PP_{P}, PU_{DC}\}\}$.
- Distribution center – Retail outlet. $Int_{DC,RO}^{role} = \{\{PP_{DC}, PU_{RO}\}\}$.

Activity 5. Our analysis showed that no coordination interaction is captured in the model.

An illustration of how the model by van der Vorst et al. is represented in our framework is presented in Figure 4.8. It should be noted that transportation and transport planning was discussed by van der Vorst et al. in such a general way that we chose not to include any transport related roles and responsibilities in our analysis. Transportation of products is implied, but it is not an important and central part of the model.

4.5.5 Dynamic Simulation of a Short Life Cycle Product Supply Chain

Finally, we present our analysis of a model by Higuchi and Troutt (2004) for simulation of a short life cycle product supply chain. By applying system dynamics-based modeling

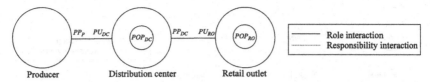

Figure 4.8: Illustration of how the model by Vorst et al. is represented in our framework.

Higuchi and Troutt studied a number of supply chain problems, such as the bullwhip effect and phantom demand. In a case study they analyzed the famous Tamagotchi case, which is a typical example of a product with a short life cycle. When the Tamagotchi, which was a virtual pet game, was launched it gained unexpected popularity and the demand widely exceeded the supply. When the manufacturing capacity was increased to meet the high demand, the demand suddenly dropped, which resulted in a large number of unsold items and a major financial loss.

Activity 1. In the model, the supply chain was divided into factory level, retail level and market level. In the analysis we chose to represent the three levels as separate organization types F, R, and M, even though it could be considered artificial to model the market as an organization type.

Activity 2. The following roles are captured in the three identified organization types:

- Factory level. $Rol_F = \{PP_F\}$.
- Retail level. $Rol_R = \{PP_R, PU_R\}$.
- Market level. $Rol_M = \{PU_M\}$.

Activity 3. Our analysis suggested that the following responsibilities are captured in the model:

- Factory level. $Resp_F = \{PRP_F\}$.
- Retail level. $Resp_R = \{POP_R\}$.

Since it could be considered artificial to represent the market as an organization we chose to not include any responsibilities in M.

Activity 4. We identified the following connections between roles:

- Factory level – Retail level. $Int_{F,R}^{role} = \{\{PP_F, PU_R\}\}$.
- Retail level – Market level. $Int_{M,R}^{role} = \{\{PP_R, PU_M\}\}$.

Activity 5. Since planning is not explicitly modeled, we found it difficult to determine how responsibilities are coordinated. Our analysis, therefore suggested that no coordination between responsibilities is represented in the model.

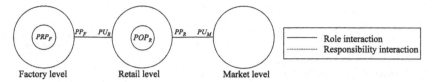

Figure 4.9: Illustration of how the model by Higuchi and Troutt is represented in our framework.

A graphical illustration of how the model by Higuchi and Troutt is represented in our framework is given in Figure 4.9. Transportation is not explicitly captured in the model, which is why we did not consider transport related roles and responsibilities in our analysis.

4.6 Concluding Remarks and Future Work

We have suggested a role-based method for structured and uniform analysis of supply chain simulation models. The method is based on a framework of supply chain roles, responsibilities, and interactions, which has been developed as a part of our work. The framework is an integral part of the method since it can be used to describe different types of organizations involved in the processes of providing and using products and transport services.

By analyzing five different supply chain models, we have partially shown the applicability and validity of the method and the framework. Since the method is based on concepts commonly used to describe agents, it was expected to perform better when analyzing agent-based models, in which planning activities and interaction can be naturally simulated. However, we have illustrated how it can also be used for analysis of simulation models that do not explicitly capture planning and interactions. This was the case in the models by van der Vorst et al. (2000), which are based on discrete event simulation, and by Higuchi and Troutt (2004), which use a system dynamics approach. This shows that our analysis method has a rather general scope, and we conclude that it can be used for structured and uniform analysis, as well as comparison, of models that are based on different modeling techniques.

When using our analysis method, sometimes the architecture of a model is analyzed and sometimes the modeled scenario is analyzed. The former often applies to models that are built as reusable frameworks for modeling of different supply chain scenarios (e.g., the models by Holmgren et al. (2012) and Strader et al. (1998)), and the latter often applies to models that are built for particular supply chain scenarios (e.g., the models by van der Vorst et al. (2000) and Higuchi and Troutt (2004)). Further, some models explicitly capture roles, responsibilities, and interactions, whereas other models implicitly capture these concepts. Therefore, it is important to emphasize that the outcome of an analysis has to be interpreted on the basis of the particular characteristics of the models that are analyzed. The outcome of an analysis is also influenced by different choices made by the analyst. Therefore, it should be emphasized that the method provides support for conducting structured analysis, but it is the responsibility of the analyst to properly use it, and to take well-motivated decisions during the analysis regarding how roles, responsibilities, and interactions are represented.

In addition to analyzing supply chain simulation models, we argue that our method can be used for analyzing real-world supply chains. For example, we believe it could be used to facilitate the process of finding the requirements of a supply chain simulation model, by identifying organizations, roles, responsibilities, and interactions that either implicitly or explicitly need to be captured in order to accurately represent a real-world supply chain.

Regarding the five models that were analyzed, we experienced that transportation often is assumed to take place, but it is typically not explicitly modeled. We consider this to be an interesting observation even though we are aware that our sample of supply chain simulation models may not be generalizable. Also, we realized that it is often difficult, or impossible, to figure out what decisions are captured in a modeled supply chain organization. Therefore, it is often possible only to roughly identify which responsibilities are captured in a model. Further, we found that it is often difficult to analyze coordination, for example, since it is typically not clearly explained how different activities and decisions are coordinated in a model. A conclusion, which is supported by the fact that roles and responsibilities often are more concrete than coordination, is that, in order to facilitate the understanding of a model, it could be less important to identify coordination than identifying roles and responsibilities.

Finally, we have contributed a framework of supply chain roles, responsibilities, and interactions that was used as the basis for an agent-based method for analyzing supply chain simulation models. A possible direction for future work is to identify additional areas of usage for the framework. The natural step would be to use the framework to define a domain-specific methodology for designing agent systems in the supply chain domain. It is believed that this could be done by coupling the framework with an already existing agent-based development method, such as Gaia (Zambonelli et al., 2003). Another direction for future work is to further develop the framework. It could be relevant to include more detail, for example, by identifying more specific roles and representing interaction by making use of speech acts. However, we have chosen to keep the framework on an abstract level since this makes it more general and also easier to apply. Also, supply chain simulation models are often not described in much detail, so it is far from obvious that making the framework more detailed would enable improved analyses.

References

Bresciani, P., Perini, A., Giorgini, P., Giunchiglia, F., Mylopoulos, J., 2004. Tropos: An agent-oriented software development methodology. AAMAS 8 (3), 203–236.

Chatfield, D.C., Hayya, J.C., Harrison, T.P., 2007. A multi-formalism architecture for agent-based, order-centric supply chain simulation. Simul. Model. Prac. Th. 15 (2), 153–175.

Chow, J.Y.J., Yang, C.H., Regan, A.C., 2010. State-of-the art of freight forecast modeling: lessons learned and the road ahead. Transportation 37 (6), 1011–1030.

Council of Supply Chain Management Professionals (CSCMP), 2013. Supply chain management terms and glossary. (Updated August 2013, http://cscmp.org/). Last accessed on Nov. 26, 2014.

Daugherty, P.J., Myers, M.B., Autry, C.W., 1999. Automatic replenishment programs: an empirical examination. J. Bus. Logist. 20 (2), 63–82.

Davidsson, P., Holmgren, J., Kyhlbäck, H., Mengistu, D., Persson, M., 2007. Applications of multi- agent- based simulation. In: Antunes, L., Takadama, K. (Eds.), Multi-Agent-Based Simulation VII. Volume 4442 of Lecture Notes in Computer Sciences. Springer, pp. 15–27.

Gjerdrum, J., Shah, N., Papageorgiou, L.G., 2001. A combined optimization and agent-based approach to supply chain modelling and performance assessment. Prod. Plan. Control 12 (1), 81–88.

Govindu, R., Chinnam, R.B., 2007. MASCF: A generic process-centered methodological framework for analysis and design of multi-agent supply chain systems. Comp. Ind. Eng. 53 (4), 584–609.

Higuchi, T., Troutt, M.D., 2004. Dynamic simulation of the supply chain for a short life cycle product – lessons from the Tamagotchi case. Comp. Oper. Res. 31 (7), 1097–1114.

Holmgren, J., Davidsson, P., Persson, J.A., Ramstedt, L., 2012. TAPAS: A multi-agent-based model for simulation of transport chains. Simul. Model. Prac. Th. 23, 1–18.

Minegishi, S., Thiel, D., 2000. System dynamics modeling and simulation of a particular food supply chain. Simul. Prac. Th. 8 (5), 321–339.

Özbayrak, M., Papadopoulou, T.C., Akgun, M., 2007. Systems dynamics modelling of a manufacturing supply chain system. Simul. Model. Prac. Th. 15 (10), 1338–1355.

Padgham, L., Winikoff, M., 2004. Developing intelligent agent systems: a practical guide. John Wiley & Sons, Chippenham, Wiltshire, England.

Persson, F., Olhager, J., 2002. Performance simulation of supply chain designs. Int. J. Prod. Econ. 77 (11), 231–245.

Roorda, M.J., Cavalcante, R., McCabe, S., Kwan, H., 2010. A conceptual framework for agent-based modelling of logistics services. Transport. Res.E Logist. Transport. Rev. 46 (1), 18–31.

Strader, T., Lin, F., Shaw, M., 1998. Simulation of order fulfillment in divergent assembly supply chains. JASSS 1 (2).

Swaminathan, J.M., Smith, S.F., Sadeh, N.M., 1998. Modeling supply chain dynamics: a multi-agent approach. Decision Sci. 29 (3), 607–632.

Tarokh, M., Golkar, M., 2006. Supply chain simulation methods. 2006 IEEE International Conference on Service Operations and Logistics, and Informatics (SOLI '06). Shanghai, China. June 21–23, 2006. pp. 448–454.

Terzi, S., Cavalieri, S., 2004. Simulation in the supply chain context: a survey. Comp. Ind. 53 (1), 3–16.

van der Vorst, J.G.A.J., Beulens, A.J.M., van Beek, P., 2000. Modelling and simulating multi-echelon food systems. Eur. J. Oper. Res. 122 (2), 354–366.

van der Zee, D., van der Vorst, J., 2005. A modeling framework for supply chain simulation: opportunities for improved decision making. Decision Sci. 36 (1), 65–95.

Wilson, R., 1934. A scientific routine for stock control. Harvard Bus Rev. 13 (1), 116–129.

Wooldridge, M.J., Jennings, N.R., 1995. Intelligent agents: theory and practice. Knowledge Eng. Rev. 10 (2), 115–152.

Zambonelli, F., Jennings, N.R., Wooldridge, M., 2003. Developing multi-agent systems: the Gaia methodology. ACM T. Softw. Eng. Meth. 12 (3), 317–370.

CHAPTER 5

Applying Delegate MAS Patterns in Designing Solutions for Dynamic Pickup and Delivery Problems

Shaza Hanif, Tom Holvoet

Department of Computer Science, KU Leuven

5.1 Introduction

Pickup and Delivery Problems (PDPs) have received considerable research attention mainly due to an increase in motorization, urbanization, and globalization. In PDPs, goods (objects or people) have to be transported from the origins to the destinations while complying with a set of constraints. With the difference in constraints and business requirements, more than 30 major variants of PDP have been identified and studied (Parragh et al., 2008).

Many approaches have been proposed and studied to deal with these variants. Some of them use combinatorial optimization approaches to calculate the best schedule in a centralized way. Such approaches require accurate global information to be managed at a centralized location. In practice, PDPs are dynamic; transportation requests appear nondeterministically; road conditions and the number of available trucks may change at runtime. This dynamism makes it very hard for the traditional one-shot planning approaches to provide an efficient and up-to-date schedule while maintaining global information for the whole system. Additionally, the real-world PDPs are large scale; a typical PDP involves hundreds of trucks and thousands of transportation requests. This means the centralized solution has to confront the problem of scalability. Due to both issues, decentralized solutions have gained more and more momentum in recent years.

Multi-agent systems (MAS) are extensively used in building decentralized solutions (Wooldridge, 2002). In these solutions, problem entities – such as trucks and packages – are modeled as autonomous and collaborative agents. Because of their decentralized nature, MAS-based solutions tend to not need centralized data/control and can provide opportunities for scalable solutions. However, engineering an MAS-based solution for PDP is known to be quite challenging. This is due to the fact that typically such decentralized solutions consist of a large number of agents that interact and cooperatively reach the system objectives.

Advances in Artificial Transportation Systems and Simulation.
Copyright © 2015 Zhejiang University Press Co., Ltd. Published by Elsevier Inc. All rights reserved.

Designing and implementing the interactions and coordination mechanisms between the agents on such a large scale, in the context of a dynamic system, is a complex task.

Many decentralized coordination mechanisms are proposed for tackling the agent interaction and coordination problem to build a decentralized MAS. These include digital pheromones (Brueckner, 2000), gradient fields (Mamei et al., 2004), market-based control (Clearwater, 1996), and tokens (Xu et al., 2005). These mechanisms are typically the result of smart engineering that combines good technical ideas with expert domain knowledge. Although such mechanisms can be exploited to design flexible solutions, a fundamental problem of these mechanisms is the lack of guidance on how to systematically choose, use, and adapt such mechanisms for a new application setting – such as a new PDP variant. Currently, all the existing knowledge and the best practices for the coordination mechanisms is spread over hundreds of papers without a clearly structured and directly usable description of the mechanisms.

Lack of this knowledge makes the design of an MAS-based solution very complex and costly (Wolf and Holvoet, 2007). In order to enable these coordination mechanisms to be used beyond a single application, they should be described in a generic, reusable fashion. Patterns are one of the most appreciated instruments for reuse in software engineering. As pointed out by Shaw and Clements (2006) "Mature engineering disciplines are characterized by reference materials that give engineers access to the fields systematic knowledge. Cataloguing architectural patterns is a first step in this direction". Patterns emerge from frequent use and experience. They identify a generic problem and provide a suitable generic solution scheme. Pattern-based design is increasingly used to construct the complex agent interaction behaviors in MAS (Aridor and Lange, 1998; Oluyomi et al., 2007; Wolf and Holvoet, 2007).

Delegate MAS patterns (Wolf and Holvoet, 2007; Holvoet et al., 2009) are a collection of interaction patterns proposed as a result of research for so-called "coordination-and-control" applications (Holvoet and Valckenaers, 2007). These patterns use the Separation of Concerns (Dijkstra, 1982) design principle by delegating part of the coordination behavior to a dedicated behavior module. They can effectively alleviate the design and implementation complexity of agent interactions. For simplicity, we use the term "Delegate MAS Patterns" to refer to this set of patterns.

In this chapter, rather than discussing and evaluating one or more MAS-based solutions to PDP variants, we illustrate how clearly defined patterns can help build such solutions. To that end, after a brief description of patterns, we first performed a core architectural breakdown of PDP to specify the elements that can be reused for building solutions of multiple PDP variants. We also identify generic challenges in designing the decentralized coordination for such MAS-based solutions. Later, the delegate MAS patterns are applied to design the agent interactions for two distinct variants of PDPs, namely the "personal rapid transportation" (Tuts, 2010) and the "hierarchical PDP" (Claes et al., 2010). We implemented our solutions in the simulation tool – MAS-DisCoSim (Gompel et al., 2010) developed in our previous work.

Although these two PDPs exhibit distinguished features and different business requirements, our experience shows that the Delegate MAS pattern can be used to effectively construct decentralized coordination mechanisms for both. This enables developers to focus on the development of the core functionality of an agent, directly related to the business requirements of a specific PDP variant. Thus, agent design and implementation complexity can be significantly reduced. This chapter focuses on the pattern-based software design. The solutions themselves for the two PDP variants can be found in our previous work (Claes et al., 2010; Tuts, 2010).

This chapter is organized as follows. We first present the related research on solving the PDPs and on patterns for MAS in Section 5.2. Then, in Section 5.3 we present a set of patterns, called "Delegate MAS patterns." We present two PDP case studies in Section 5.4, and discuss solutions for both case studies in Section 5.5 – these solutions make use of the presented patterns, and as such can reuse the know-how and the quality expectations that the patterns aim to accomplish. We conclude our work in Section 5.6.

5.2 Related Work

In the last decade, various taxonomical papers have appeared (Parragh et al., 2008; Berbeglia et al., 2010) which classify the diverse approaches proposed for solving different variants of PDP. Here, we differentiate two major streams of approaches: (1) combinatorial optimization-based approaches and (2) MAS-based approaches. Finally, we present related work with respect to the pattern-based MAS design.

5.2.1 Combinatorial Optimization-Based Approaches for Solving PDP

Traditional approaches mainly treat PDPs as static problems and solve different variants using the combinatorial optimization techniques (Burke and Kendall, 2005). These approaches are not able to effectively handle the dynamism in the problem (Parragh et al., 2008). In order to deal with the dynamic problems, online algorithms (Berbeglia et al., 2010) are designed which add an innovative heuristic on top of the static solutions. When new data becomes available, instead of recomputing the whole solution, they reoptimize the existing one using heuristics and metaheuristics (Gendreau et al., 2006; Gutenschwager et al., 2004).

Moreover, in both the combinatorial optimization and the reoptimization based approaches, the role of a centralized dispatcher is evident. They assume that global information is maintained at the dispatcher, and calculate optimal routes for serving the set of requests available in a certain amount of time. As centralized approaches, reoptimization-based approaches face the challenge of rapidly responding to new information as it becomes available.

In order to meet the challenge of scalability, the idea of breaking the problem into component parts appeared. Ghiani et al. (2003) review parallel computing solutions for solving dynamic

PDP. Such solutions use different control and communication structures (e.g., master-slave, etc.) to efficiently search the solution space on the basis of parallel computing. However, they overlook the possibilities of modeling the system in a decentralized way. For example, PDP could be reframed on the basis of different entities within the problem, that is, the transportation requests and the vehicles. The following section describes the decentralized, MAS-based approaches for PDP.

5.2.2 MAS-Based Approaches for Solving PDP

MAS present a modeling abstraction for building decentralized solutions (Jennings, 2000), with the ability to perform well in uncertain domains (Fischer et al., 1995). Fischer et al. argue that MAS fit the transportation domain particularly well (Fischer et al., 1995). The main motivation of this statement is the following. First, the transportation domain is inherently decentralized (vehicles, transportation requests, companies, etc.). Second, decentralized MAS can handle the dynamism more appropriately. Third, commercial companies may be reluctant to provide proprietary data needed for global optimization whereas agents can use local information.

To support their claims, Fischer et al. present an MAS-based solution in which tasks are allocated to trucks and company agents using an extension of the Contract Net Protocol (CNP). In their approach, the CNP is used to calculate a basic solution, which is later improved by using auction protocols (Wooldridge, 2002). In essence, the authors recognize the limitation of a fully decentralized system, that is, that agents only have access to local information; and hence they make agents coordinate using a two-level approach. The company agent at the top of the hierarchy has more information about the system and makes decisions about the individual truck agents that have limited information. Mes et al. (2007) and Dorer and Calisti (2005) also try to find a balance between the omniscience of a centralized approach and the swiftness of decentralized MAS. Mes et al. (2007) use two high-level agents – the planner and the customer agent – to collect information from and distribute information to the agents under their responsibility. Dorer and Calisti (2005) use a centralized dispatcher agent to allocate incoming requests to regional dispatcher agents. In both approaches, the role of the higher-level agents is to centralize the information essential for the lower-level agents to make effective and more optimized decisions.

Beside such basically market-inspired techniques (Clearwater, 1996), other decentralized coordination mechanisms are proposed, including digital pheromones (Brueckner, 2000), gradient fields (Mamei et al., 2004), and tokens (Xu et al., 2005).

In essence, these solutions try to design smart engineering techniques to provide an optimized solution for a particular PDP variant. However, due to the differences in constraints and business requirements, the tricky balance achieved in one solution for a PDP variant normally could not work in the other variants. This makes the solutions discussed here very hard to be reused. For a new PDP variant, having different business requirements, a new solution is required to be designed. Lack of reusable design knowledge makes the MAS-based PDP solutions very hard to engineer.

This problem results from lack of reusable assets as well as guidance on how to systematically choose and use the most suitable coordination mechanism for MAS. In order to deal with this limitation, some researchers proposed using reusable patterns for designing decentralized MAS.

5.2.3 Patterns for MAS

Aridor and Lange (1998) are among the pioneers who capture MAS experiences as patterns. They identify the importance of structuring coordination mechanisms and present mobile agent patterns under the three classes of Traveling, Task, and Interaction patterns. Their Message pattern (in the class of Interaction patterns) provides the foundation for our Smart Message pattern discussed in Section 5.3.1.

More recently, Oluyomi et al. (2007) presented a framework to facilitate the classification, analysis, and description of agent-oriented patterns. Their primary concern is to present patterns for covering all levels of agent-oriented software engineering but they do not demonstrate the applicability of patterns with fully fledged case studies.

In our previous work, Wolf and Holvoet (2007) distil patterns from the recurrent solutions for the decentralized coordination between agents. They presented the Gradient Field pattern and Market-Based Control pattern. Later, the Delegate MAS pattern (Holvoet et al. 2009) was proposed to build complex interactions behaviors for a set of coordination-and-control applications. It provides an extensible and tunable solution to balance global information and local decisions discussed in Section 5.2.2.

Although dynamic PDP variants exhibit significant differences in their constraints and the business requirements, they still have many common requirements. For instance, there are always vehicles and customer requests in a dynamic and large-scale environment. Having said that, patterns should be able to play an important role in the solution design. However, there is little effort to demonstrate the (re)use of patterns in designing different variants of PDP.

5.3 Delegate MAS Patterns

A literature study on coordination mechanisms for decentralized applications revealed many recurring techniques for typical technical challenges. One line of techniques has been captured as a set of patterns in (Holvoet et al., 2009). In this section, we describe two of these patterns, called Delegate MAS patterns. The patterns are described using a typical template for patterns, including a description of the context, problem, force, solution, and consequences.

This section first introduces the smart message pattern, which is used for communication between agents in changing environments. Then, the Delegate MAS pattern is described for managing such smart messages (Section 5.3.2). For a clear description of concepts, in this section, we will use the vehicle agent and the package agent that are explained in Section 5.5.1.

5.3.1 Smart Message

Context

In large-scale dynamic systems, due to many dynamic factors, the environment of a distributed system normally exhibits dynamic features. In such systems, the environment is represented as a graph in which the nodes have connections to the neighboring nodes with varying quality of service and changing connections. Application-level software entities need to interact with other entities on top of this changing topology for information distribution, information gathering, coordination, and synchronization.

Problem

In general, a problem occurs when multiple and complex interactions are required between one entity and other entities. In MAS, these entities are normally implemented as agents. Under various circumstances, it is not possible or not desirable to have simple direct interactions with an agent. This situation happens when there is no accurate global information available, for instance, if the exact identification or the location of the other agent is not known, or a route to the other agent is not known/broken. Using the traditional mechanism to send messages back and forth between the two remote nodes can be troublesome.

Forces

The dynamic nature of the decentralized systems makes the communication between different agents very hard to engineer. A solution needs to cope with the dynamic topological and quality of service characteristics while communicating with other agents. At the same time, heavy communication between distant nodes needs to be limited to avoid the excessive overhead.

In order to provide all its interactions in an uncertain environment, an agent needs to be developed covering both its functional logic and the nonfunctional communication requirements, while overcoming the previous mentioned problems. This dual responsibility makes the design of such an agent very complex and costly. A solution is needed to alleviate or manage this complexity.

Solution

The pattern identifies the smart message as the core building block to solve the mentioned problem – see Figure 5.1. A smart message is a self-contained entity that comprises both state

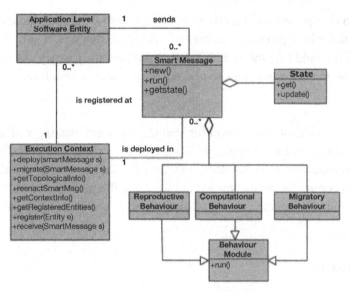

Figure 5.1: A smart message

and behavior, and retains information about the agent that is responsible for the message (e.g., the sender). The smart message autonomously moves in the environment and interacts with nodes on behalf of its owner agent. Behavior of a smart message typically includes:

- querying context information, and possibly interacting with locally active agents;
- local computation that complies with the messages' objective;
- migratory behavior that decides which node to move to next;
- reproductive behavior that allows the new smart messages or clones of itself to be created and spawned in the execution context.

A smart message is deployed in an execution context, that is, an environment in which the smart message can perform its behavior and update its state. This execution context normally links to a node with computing capability. The execution context offers services for:

- creating new smart messages;
- migrating a smart message to a neighboring node;
- receiving and consequently reenacting smart messages (i.e., triggering their behavior execution);
- querying for the context information, including for topological information (e.g., providing lists of neighboring nodes) and for the other registered entities (e.g., providing references to the agents that are active at the node).

In the PDP, as will be discussed further below, the vehicle agent can send out a smart message to explore nearby packages and evaluate the quality of various paths that the vehicle could make. It can use a disciplined flood (Holvoet et al., 2009) to limit communication overhead in its search, combined with a property itinerary of several destinations it aims to visit. In order to assess the time it would take the vehicle itself to perform the journey, the smart message can also interrogate the nodes it passes to find out about their current/forecast road conditions.

Consequences

A smart message is a valuable instrument for agents for communicating with each other in a large-scale and unknown environment, which is typical in dynamic PDPs. Care should be taken that their reproductive behavior does not cause flooding or eternal traveling behavior. Due to the limited communication capacity, the state and the mobile code must be limited in size. Due to the limited computational capacities of nodes, the behavior must be of limited complexity.

5.3.2 Delegate MAS

A smart message is a single, mobile unit. The use of a smart message can alleviate the complexity in implementing communication on top of a changing environment. However, the smart messages can be effectively used in a conglomerate way, collectively executing a particular task or role. A Delegate MAS pattern is designed to provide such management.

Context

Similarly, as for the smart messages, the problem that is addressed by this pattern arises in the context of applications in large scale, dynamic decentralized systems. The underlying, distributed communication environment is a (dynamic) graph topology, where nodes have connections to the neighboring nodes, possibly with varying quality of service. Because of dynamism, agents – such as the package agents and vehicle agents – require updated information. They need to interact repeatedly over a period of time with other agents, for coordination, synchronization, and information gathering.

Problem

The repeated interactions with various agents need to be managed appropriately, which includes the specification of the individual interactions, the frequency of interactions, the aggregation, and the processing of the results from the interactions. A software design is needed to make efficient and effective use of the smart messages, without making the implementation of the agent too complex.

Forces

A set of related interactions, collectively pursuing an objective of an agent, needs to be managed in order to ensure coherence of their behavior while avoiding unnecessary overheads. These related interactions (made using smart messages) should be performed

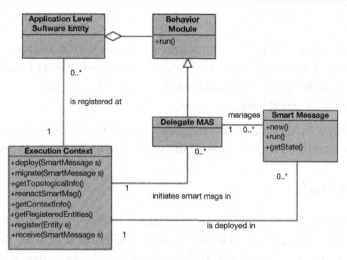

Figure 5.2: Delegate MAS

coherently, according to a shared policy. For example, one agent that needs to interact with five distant nodes in order to find out which of the nodes can be reached the fastest. It must ensure that the individual interactions with the nodes rely on the same objectives and evaluation criteria, and the information that the interactions produce must be interpreted in a coherent manner.

Communication between the distant nodes should be limited as the overhead of such communication can be substantial. For example, interactions to continuously monitor a path between two nodes should not flood the network.

Solution

A Delegate MAS (see Figure 5.2) is a behavior module (Maes, 1990), that is, a well-defined behavior that an agent can perform to reach a particular objective or task. The behavior module is monitored and controlled by an agent, and fulfils a well-defined objective or task on behalf of the agent. An agent's behavior consists of selecting and executing behavior modules, possibly in a concurrent manner. The agent itself manages the activation and deactivation of the behavior modules, as well as the coordination between behavior modules.

A Delegate MAS uses smart messages to fulfill its objective or task. As such, a delegate MAS is in charge of the management of the smart messages, and encapsulates a policy for creating smart messages (including a policy about timing and frequency of creating messages) with their own suitable (reconfigurable) behavior and initial state, and spawning the messages through the execution context of the node. Additionally, the Delegate MAS module collects the results that smart messages report back. The results are preprocessed according the specification of the behavior module. Later these results are forwarded to the agent for further interpretation.

Consequences

This solution allows agents to delegate part of their communication and interaction behavior to a separate module. Using this pattern requires careful consideration of which aspects of the decision-making can be fully delegated to this behavior module and which aspects must remain controlled by the agent itself.

By using the Delegate MAS pattern, an agent can delegate the task of interactions to the separate behavior module so that the complexity of the agent can be simplified.

5.4 Two Case Studies

In this section, two different PDPs are introduced: the "personal rapid transportation" (Tuts, 2010) and the "hierarchical pickup and delivery" (Claes et al., 2010). These two variants have different business requirements and system objectives. However, they face similar coordination-and-control problems that make agents' implementation complex and error prone.

5.4.1 Personal Rapid Transportation

A Personal Rapid Transportation (PRT) system is a public transportation model featuring small vehicles to provide personalized transportation services to the customers. It consists of a number of pick-up/drop-off points (called stations) scattered in the service area. Customers at the PRT stations wait to be picked up by the vehicles and delivered to his/her destination station with certain time constraints on both waiting and delivery time. A number of PRT vehicles drive around the service area and stop at certain stations to pick up, transport, and deliver customers.

Compared to other public transportation systems, PRT has some following distinct features:

- PRT vehicles do not stop at every station. They are designed to provide a nonstop journey for the customers.
- The routes of PRT vehicles are not fixed. They can autonomously change their routes according to the current conditions.
- PRT vehicles normally have smaller capacity compared to the normal vehicles.

Figure 5.3 shows a simulated PRT for Leuven City (retrieved from MAS-DisCoSim (Gompel et al., 2010)). In this figure, there are 21 PRT stations marked with dots and 7 vehicles to pick up customers. The customers are shown in the picture with package icons.

The general goal is to deliver all customers while maximizing a function of customer satisfaction and minimizing unnecessary travel of the vehicles.

5.4.2 Hierarchical Pickup and Delivery

Large logistics providers organize their transportation network in a hierarchical "hub and spoke" overlay network (Rodrigue et al., 2006). In order to reach their destination, the

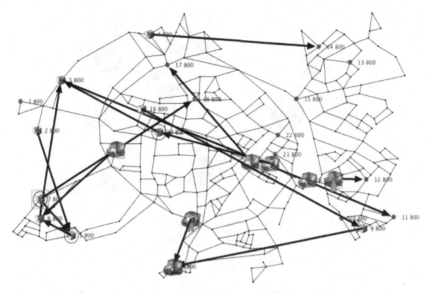

Figure 5.3: Personal Rapid Transport (Leuven Map)

packages are routed from depot to depot in this overlay network. If the destination of a package falls outside the region of a particular depot, it is forwarded to a depot on a higher level. This process is repeated until the package reaches a depot covering its destination or the highest level of the hierarchy is reached. At the highest level, the top-level hubs are connected in a mesh-like fashion. An example of the hierarchical hub-and-spoke overlay network is depicted in Figure 5.4.

The packages need to be routed through this hierarchical network in order to reach their destination. Between each level in the hierarchy, the transportation resources (e.g., trucks, trains, ships, aeroplanes) need to be allocated in order to actually transport the packages up and down the hierarchy. Scheduling these resources is a difficult process due to a number of challenges, including the dynamic nature of the transportation demand, the scale of the hierarchical overlay network, and unforeseen events – for example, the eruption of the Eyjafjallajökull volcano in 2010 – that disrupted global transportation, severely influencing the schedules.

The general goal is to deliver all the packages, while minimizing the resource usage. As the packages move up in the hierarchy, they need to be grouped together to save the resource usage.

5.5 Solutions

This section presents solutions for both PDP variants discussed above. This section, however, does not claim nor aim to present a detailed quantitative evaluation of these solutions. Our main objective is to illustrate the application of the Delegate MAS patterns for engineering valuable solutions.

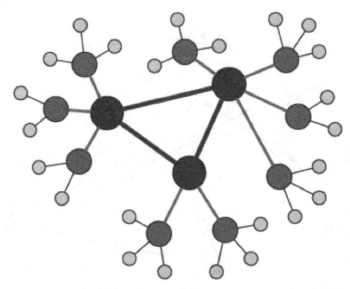

Figure 5.4: Hierarchical PDP, largest circles represent top level of hierarchy

Both solutions start from a core architectural breakdown that is common in MAS-based solutions for PDPs (see Section 5.2), and with a description of common technical challenges. Thereafter, we refine the solutions using the Delegate MAS patterns.

5.5.1 Core Architectural Breakdown

PDPs consist of the transportation requests, which are handled by the transport devices (vehicles). As indicated by Fischer et al., 1995, PDP is an inherently distributed problem, and to model the problem entities, such as vehicles and transportation requests, as agents is a natural metaphor (Fischer et al., 1995). Therefore, in our core architectural breakdown, the transportation requests are modeled as package agents and the transport devices are modeled as vehicle agents, both being deployed in a common environment.

The environment of the targeted PDP is modeled as a directed graph. This graph represents a (simplified) road structure, where the nodes represent physical or virtual road connections. The packages (nonmoving static entities) are deployed in the environment on the nodes. Nodes have storage capability and agents can store/acquire information from nodes. In order to deliver the packages, the mobile entities – vehicles – move from one location to the other location in the environment. To enable communication through the environment, both the packages and the vehicles can communicate with the environment. When a vehicle reaches a particular package, it can perform an action (pickup or enquiry) on it.

Agents

As mentioned earlier, the problem entities that represent the autonomous units of decision-making, are modeled as agents. An agent is deployed on a physical entity, which in turn is situated in the environment. It contains not only entity specific information but also the processing logic related to it. The physical entity can be abstractly described as a set of capabilities, where a capability specifies the operations that the entity is able to perform. For instance, a vehicle has the capability to move, pickup, and deliver packages; both package agents and vehicle agents have the capability to communicate with other agents.

In accordance with the business requirements, an agent is responsible for performing certain tasks correctly (e.g., a vehicle should deliver at the right location) and timely (e.g., a vehicle should pickup packages at the specified time). It needs to explore and plan to achieve its individual goals, which is preferably coherent with the system objectives. For instance, the vehicle agent is responsible for performing its task by guiding its vehicle through the environment, and by communicating and coordinating with the other agents.

Modeling and implementing the internal behavior of the vehicle or package agents is another crucial architectural decision. Various agent architectures can be used, for example, a BDI (Belief–Desire–Intention) architecture (Rao and Georgeff, 1995), or a Reactive Agent Model (Wooldridge, 2002). Again, this chapter does not pretend to "per se" provide the best solution for both PDP variants, but illustrates the application of patterns. For our solutions, we have opted to use a BDI-based architecture for the agents. Its explicit notion of beliefs, desires, and intentions suits the way we modeled the problem.

Agent interactions

The agents are deployed on the physical entities in the environment, and obviously need to interact to achieve the system objectives. A typical interaction amongst these agents could be as follows:

When a new transportation request – that is, a package – enters the system, a package agent is created. The agent is aware of the package details, that is, the pickup and delivery location, the time to pickup, its weight, and so forth. It can respond to the other agents querying this information. Vehicle agents are aware of the current position of their vehicle, and can steer their vehicle to move in the environment and pickup/deliver the packages. It observes the packages in its vicinity and inquires the package details from these package agents. On the basis of its decision logic, it considers the possible and feasible packages to pickup, then selects a particular package and communicates its intention (to pickup) with the package agent by sending a proposal to the package agent.

A package agent in turn can choose to commit to the most suitable proposal offered by various vehicle agents. This allows agents to coordinate their behavior by accommodating their intentions. Coordination is necessary as actions of one vehicle agent may influence the beliefs (and in turn plans) of the other vehicle agents. If the situation in the environment is such that picking up another package becomes substantially more favorable (depending on business requirements), the vehicle agent can reconsider its reservation and adopt a new intention. This process is called intention reconsideration (Wooldridge, 2002).

In this core architectural breakdown, these agent interactions are decentralized and localized, that is, they do not require accessing any global information for coordinating their actions. This feature enables the approach to be scalable – in terms of number of vehicles, number of transportation requests, or geographical expansion. Moreover, agents are able to respond quickly to the local changes in the dynamic environment.

5.5.2 Challenges in Designing MAS-Based Solutions for PDP

There are several distinguishing characteristics of PDP that make the designing of a solution – and, in particular, the interactions between agents – challenging. These challenges are faced in the context of a dynamic and scalable environment. The environment is dynamic since traffic jams, road constructions, vehicle failures, dynamically arriving transportation requests, request cancellations are the causes of uncertainty in the environment. The problem is large in scale – in terms of number of vehicles, types of vehicles (single load carriers, trolley trucks, trains, airplanes, ships), number of transportation requests or widely distributed locations of pickup or delivery.

We describe the most important challenges to address:

1. The resources are limited. Any agent, such as a truck agent, cannot maintain accurate global information due to limited capacity and communication capability. At the same time, the computational resources are limited as well. Similarly, there are other, physical constraints, such as a vehicle's speed range, road capacity, and so forth.
2. The decisions are highly correlated. The decision of one agent influences the behavior of the other agents. When a vehicle finds a more suitable package to pick from the one it reserved, it may cancel the previous reservation and go for a new reservation. This cancellation may cause other vehicle agents to propose the package for the pickup. This implies that an agent needs to notify other agents about its intention and requires the other agents to keep track of resources it has reserved.
3. Purely local decisions are not optimal. If the vehicle and the package agents make decisions purely on the basis of local information, the resulting solution will be far from optimal. On the other hand, as indicated in Section 5.2, centralized approaches tend to breakdown when the underlying system is large scale and agents are unable to give an agile response to the local changes. It is challenging to maintain a decentralized design yet enable agents to: (1) disseminate necessary local information to the remote agents; (2) collect the disseminated information that is necessary for local decisions.

Most of these challenges are related to the coordination between agents, making the design of collaboration crucial for MAS-based solutions, yet highly complex in general.

5.5.3 Typical Applications of Delegate MAS Patterns

We first point out the variety of possible tasks that can be accomplished by using Delegate MAS patterns by referring to the core architectural breakdown, that is, identifying the vehicle agents and the package agents. Later, we refine the solutions for both case studies.

Local-to-(sub)global information dissemination

In a large-scale, decentralized system, local decisions are not and cannot be optimal. Referring back to Delegate MAS patterns discussion in Section 5.3, an agent can only effectively make decisions when it receives the relevant information about its execution context. This means that agents need to attract relevant information toward them, and propagate their local information to the remote nodes. Through this information dissemination process, agents can maintain an updated model of (the relevant part of) the topology. In order to provide this support, agents can use feasibility delegate MAS.

Feasibility delegate MAS is an instance of a delegate MAS pattern for information dissemination and topology discovery by sending lightweight smart messages, so-called feasibility ants. The feasibility ants are issued by their master agent – for instance, a package agent – to propagate local information or changes to the remote nodes within a certain range. A feasibility ant carries out information designated by its master agent (here the package agent) and roams in the environment. At each node it passes, it drops information about the feasible path to its master agent starting from this node. Then, it clones itself for every node directly connected to its current node, except for its incoming node, until it reaches the same node already or exceeds the range limitation. During the roaming process, feasibility ants make local information (sub)globally available and maintain updated information of feasible paths. Vehicles use these feasible paths to reach the package.

Exploring feasible paths

The information spread using the feasibility ants is scattered in the environment. In order to make a decision on the basis of this updated and accurate data, an agent – such as a vehicle agent – needs to actively retrieve information in which it is interested. This challenge can be solved by using the exploration delegate MAS – an instance of Delegate MAS pattern. This instance sends exploration ants. An exploration ant is a smart message that is designed to explore a feasible path for its master agent by collecting and evaluating the information distributed by the feasibility ants. It selects a feasible path and explores it by moving node to node. It evaluates the explored path quality in terms of time, cost, and other quality criteria from its master agent's perspective. In this way, the master agent can be alleviated from the burden of doing such a decentralized and concurrent exploration task.

Negotiation between agents

In order to deal with the challenge of multiinfluential decisions, agents need better coordination by negotiating. Whenever an intention is made that might influence other agents, it is preferable to notify and negotiate with other agents about it. If negotiation succeeds, the other agent might be required to reserve certain resources. To serve this purpose, the intention delegate MAS can be used. This instance of the Delegate MAS pattern manages smart messages, the intension ants, to propagate the intention of its master agent through the environment. For instance, the vehicle agent can send intention ants to propose to a package agent its intention to pick up at a particular time and with a certain delivery cost. The package agent may select one of the best offers proposed by different vehicle agents.

The challenge of dealing with limited resources is also addressed by a Delegate MAS pattern, as the master agent only needs to deal with its core functionality and delegates most of the interaction functionality to the ant agents. Compared to the monolithic solutions where one agent implements all those functionalities, by using the Delegate MAS pattern, a master agent consumes far less resources. Moreover, no entity needs to maintain global information, rather agents disseminate and collect only their relevant information.

Dynamism

Due to the dynamism of the system, these three different kinds of ants need to be sent out periodically. After changes in the environment, old information about feasible paths, path evaluations, and reservations might become outdated. In order to achieve consistency with the changing environments, ants will terminate their execution according to "time to live" stamps assigned to those ants. Detailed design and sample implementation can be found in (Holvoet and Valckenaers, 2007).

These three different instances of a Delegate MAS pattern – feasibility ants, exploration ants, and intention ants – cover the three important requirements typical in the decentralized coordination-and-control applications. Moreover, according to application-specific requirements, other instances can be developed on the basis of the Delegate MAS patterns. In the subsection below, we will introduce how to use a Delegate MAS pattern in designing the solutions for the selected case studies.

5.5.4 Personal Rapid Transportation

Agent-based modeling

As described in the previous section, PRT is intrinsically distributed. Three problem entities can be identified: (1) customers, (2) PRT vehicles, and (3) PRT stations. In order to use an MAS-based approach to model this variant of PDP, we refine the core architectural breakdown for PDP (described in Section 5.5.1) and identify that three types of agents – (1) Customer agent, (2) Vehicle agent, and (3) Station agent – are needed. However, as a customer can only be taken up/dropped off at stations, a Customer agent is represented by the Station agent to reduce modeling and implementation complexity.

Station agent

The Station agents are basically containers for the customers. As a customer can enter the station at an arbitrary time, a Station agent needs to collect this information and propagate it to the Vehicle agents. They also help customers in vehicle selection when multiple options are available.

Vehicle agent

The major functions of a Vehicle agent are to pick up waiting customers and to deliver them to their destinations. In order to achieve this task, a Vehicle agent must collect information about the waiting customers: pickup and delivery locations and the time constraints of a specific customer, and so forth. When multiple customers are discovered, it needs to decide on which customer to take so as to maximize users' satisfaction and reduce unnecessary travel. After a decision is made, this agent should guide the vehicle to the pickup location within changing road conditions. Its functions also include finding the best route to deliver the customer.

Coordination requirements

In PRT, we identify several features that meet the context of the Delegate MAS pattern. It is intrinsically decentralized, with many agents each having their specific goals. It also involves complex interactions between agents to achieve decentralized coordination. In order to deliver the customers in a cost-effective way, these agents must effectively communicate with each other in the changing environment.

Applying Delegate MAS patterns

In a dynamic PRT system, the major function of a Vehicle agent is to provide functions for customer searching, picking up, and delivery. However, due to the dynamic and decentralized nature of the system, an agent needs far more complex implementation.

Key functions of a Vehicle agent are listed as follows:

1. To search for the customer information in a localized way;
2. To find a possible/optimal route to reach the customer for each detected waiting customer;
3. To choose one customer by using heuristics;
4. To make a reservation (negotiates) with a Station agent;
5. To guide the vehicle to the reserved customer (station) in a changing topology.

In all these requirements, from a Vehicle agent's point of view, only the tasks 3 and 5 are its core functionalities. However, the exploration and the coordination requirements though

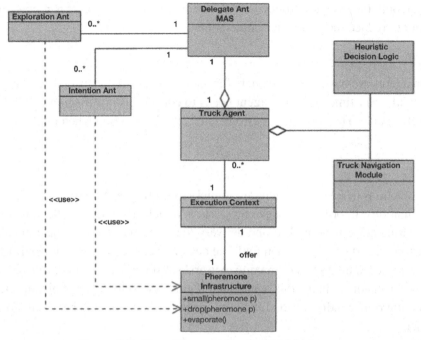

Figure 5.5: Class Diagram for the Vehicle Agent

not core functionality, constitute an important section in the agent implementation and add considerable implementation complexity.

The Delegate MAS pattern provides a systematic approach for implementing the complex interactions between the agents. Thus, this pattern can be used to simplify agent implementation. Here, we only focus on illustrating the design for the Vehicle agent. The class diagram of the Vehicle agent is shown in Figure 5.5.

Vehicle agent

A Vehicle agent uses the exploration ants to explore feasible paths. The exploration ants detect the available customers from the information propagated by the feasibility ants (task 1). It calculates the best route to the target customer on the basis of information collected during the exploration (task 2). Vehicle agents use the intention ants to notify a Station agent that they are likely to visit this station at the estimated time and with certain cost to pay for the delivery (task 4). In this way, the intention ants make a (evaporable) booking on the resources and the Station agent adjusts the status of its customers. Therefore, the Station agent is able to predict the time and the cost for the delivery more accurately. Thus, it is possible to make more refined/timely scheduling for its customers.

Table 5.1: Implementation complexity for different modules.

Functional Modules	Submodules	Line of Code	Percent (%)
Delegate MAS	Feasibility Ant	259	64.5
	Intention Ant	125	
	Exploration Ant	167	
	Delegate MAS Management	218	
Decision logics	Target selection, Route decision, negotiation	423	35.5

By using the Delegate MAS pattern, the functionality of the Vehicle agent is implemented in a simplified way. To quantify this statement, we have analyzed the code for implementing this functionality.

Table 5.1 shows that: (1) A considerable part of the agent's implementation (almost 2/3rd) can be separated from the decision logic of the agent. This implies that the Delegate MAS is able to segregate concerns, and supports the agent's design to be based on the design principle of Separation of Concerns 2). A substantial part of a Vehicle agent's implementation (about 200% of the agent's core functionality) can be explicitly supported by Delegate MAS patterns.

The Station agent was also designed using a Delegate MAS pattern by delegating the information dissemination task to the feasibility ants. It also uses the heuristic to decide which reservation for a customer should be chosen when multiple proposals from the Vehicle agents are received. Detailed discussion of this agent can be found in the reference Tuts (2010).

5.5.5 Hierarchical Pickup and Delivery

Agent-based modeling

Also for the hierarchical PDP, we describe a solution on the basis of a set of coordinating agents. Two types of agents are identified.

Package agent

Every Package agent is responsible for one package. It represents its package's information, such as the source, destination, time constraints, expense constraints, and so forth. Its major function is to find a suitable route through the hierarchical overlay network. It travels alongside the package until it reaches its destination.

Depot agent

For each depot, a Depot agent is created. The major function of this agent is to make a best schedule to allocate optimized resources to the packages among the hierarchical overlay network. This schedule has two constraints: (1) satisfy packages' delivery requirements; (2) reduce the resource usage by combining the delivery of packages as much as possible.

Due to the dynamism in the system, a Depot agent has to continually optimize its local schedule according to the historical data and the reservations proposed by the Package agents.

Compared to the core architectural breakdown, this model does not have Vehicle agents. Since, in this PDP variant, vehicles can only travel between depots and are fully regulated by Depot agents, the Depot agents can take up the functionality of the Vehicle agents.

By communicating with each other, these agents collectively decide upon the route that a package will traverse through the hierarchical overlay network. They also calculate the allocation and the schedule of the transportation resources. Like the previous solution, in this solution also no global knowledge is assumed, every node in the network only knows about its immediate neighbors up and down.

In contrast with the solution technique to the previous case study, this one is package centric, that is, it is the Package agent that needs to find suitable routes for its package. As Package agents can influence the schedule of selected Depot agents and have to compete with the other Package agents, finding a route through the overlay network and the resources necessary to transport the package become challenging.

Coordination requirements

As can be seen from the problem description, Package agents have two different interactions with the Depot agents: (1) Search in the hierarchical overlay network to find a feasible path by retrieving local schedule information from Depot agents; (2) If a suitable route is selected from all the possible explored routes, the Package agent needs to reserve the resources with the Depot agents along the selected route. These two types of interactions require the Package agents to interact with the multiple remote Depot agents in a concurrent way. These interactions meet the context of the Delegate MAS pattern as described in the next section.

Applying Delegate MAS patterns

In order to simplify agent interactions, the Delegate MAS pattern is used to design both the Package agent and the Depot agent. The Package agents can delegate the path exploration task to the exploration ants. Figure 5.6 shows an example of how the exploration ants explore the hierarchical overlay network. The second requirement for the resource reservation can be typically implemented by using the intention ants. The Depot agents are implemented in a passive way. They respond to requests rather than actively initiate interactions.

In order to deliver a package, agents need to interact with others while making local decisions. This process constitutes the following steps:

1. The exploration ants roam the hierarchical overlay network in search of feasible paths;
2. The Depot agents inform passing exploration ants about the schedules and limitations of their resources;
3. The exploration ants finish their exploration phase and gather at the package destination;

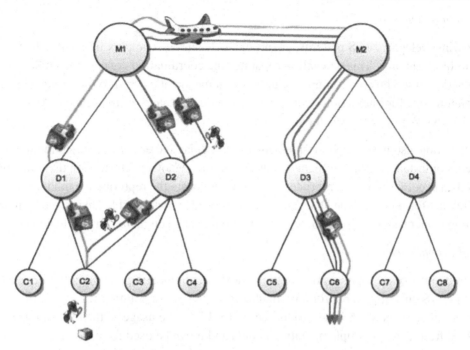

Figure 5.6: Exploration ants explore the hierarchical network.
On arrival the exploration ants have a schedule including arrival and departure times
of the transportation resources. In this figure, exploration ants search for feasible
paths between nodes C2 and C6.

4. One possible path is selected using the heuristic at destination;
   The reservation for the package schedule is carried out with the Depot agents across the
   path by an intention ant through the selected path;
5. Depot agents use the reservation information to inform the future exploration ants and
   optimize their local schedule.

In these steps, we can see the interactions play an important part in the implementation.
By using the Delegate MAS pattern, developers of Package agents can focus on the core
functionality – the heuristic for path selection. The same applies to the development of the
Depot agent, which only needs to implement the local schedule optimization. Due to this
design based on separation of concerns, implementation of this solution has similar code
features as the solution introduced in Section 5.5.4.

5.5.6 Discussion

In both case studies, we demonstrated the use of the Delegate MAS pattern in designing
decentralized solutions for dynamic PDP. Our experience shows that the Delegate MAS
pattern gives support in the following way.

Separation of concerns

By applying Delegate MAS patterns, agents can delegate the complex interaction behavior to the lightweight ants. This delegation separates the coordination concerns from the agent's core functionality. Thus programmers can focus on the agent's core programming and most of the skeleton code for interactions can be generated due to usage of the Delegate MAS pattern. Table 5.1, to some extent, gives the evidence.

This separation also makes agent implementation modular and easy to be (re)configured when required. For instance, if the problem details changes and another information dissemination mechanism becomes more appropriate, we can easily locate the reproductive module of the feasibility ant for revision. The Delegate MAS pattern also makes ant behaviors easily tunable – for example, increasing the frequency of sending out ants when there is more dynamism.

Extensible implementation

Compared to the other solutions that implement smart engineering techniques to devise complex MAS only applicable for a particular application, our solution uses reusable patterns to implement MAS based coordination for PDP. The usage of the pattern means it is abstracted from concrete implementation details and it can be used for different requirements.

In these two case studies, we used three types of instances that implemented a Delegate MAS pattern. However, the pattern itself does not restrict what exact functionalities a Delegate MAS module should perform. Different implementations can be created to accomplish application specific tasks in a decentralized and concurrent way. This means our solution can be extended to support a variety of interaction behaviors between agents.

5.6 Conclusion

Large-scale and dynamic, decentralized applications are particularly hard to engineer. This is especially true for designing MAS-based solutions for dynamic PDP. Many solutions have been described in the literature, yet they have not been consolidated into reusable assets and are often optimized for a particular application. This limitation makes them very hard to be reused for designing different PDP variants.

In this chapter, we did not focused on evaluating specific solutions. Rather our focus has been on using reusable patterns to design agent-based solutions for coordination-and-control applications considering PDP as a case study. The Delegate MAS patterns, proposed in our previous work to provide a well-designed solution for implementing decentralized agent interactions, are applied to design the solutions for two PDP variants. Delegate MAS leads to decentralized approaches, that is, solutions that do not require global information to be managed at a centralized entity for optimized decision making. Instead, agents make decisions locally by collaboratively disseminating and collecting information, and by coordinating using lightweight ants in a fully decentralized way. Our experience shows that by using these reusable patterns, an agent's internal design and development can be largely simplified. The use of the Delegate MAS pattern proves to be tailorable and extensible.

Many challenges for further work remain. On the agenda is extending the application of the Delegate MAS pattern to other variants of PDP. In this way, instances other than the three typical instances of delegate MAS can be identified. Another direction is to investigate and abstract the additional recurrent patterns from the existing solutions for PDP. A pattern repository for the PDP can be built to provide a systematic design for the PDP. It is also interesting to investigate how to effectively combine multiple patterns for a solution design. Such a pattern repository can then be studied for its applicability in other application domains that require decentralized coordination in large-scale and dynamic settings.

References

Aridor, Y., Lange, D.B., 1998. Agent design patterns: elements of agent application design presented at the Proceedings of the Second International Conference on Autonomous agents. Minneapolis, Minnesota, United States.

Berbeglia, G., Cordeau, J.-F., Laporte, G., 2010. Dynamic pickup and delivery problems. European Journal of Operational Research 202, 8–15.

Brueckner, S., 2000. Return from the ant – synthetic ecosystems for manufacturing control. PhD Thesis, Humboldt-Universität, Berlin.

Burke, E., Kendall, G., 2005. Search methodologies: introductory tutorials in optimization and decision support techniques. Springer, New York.

Claes, R., Holvoet, T., Van, G.J., 2010. Coordination in hierarchical pickup and delivery problems using Delegate multi-agent systems. In the fourth Workshop on Artificial Transportation Systems and Simulation. 1–7.

Clearwater, S.H., 1996. Market-based control: a paradigm for distributed resource allocation. River Edge, N.J.: World Scientific, Singapore.

Dijkstra, E.W., 1982. On the role of scientific thought. Selected Writings on Computing: A Personal Perspective. Springer-Verlag, New York, pp. 60-66.

Dorer, K., Calisti, K., 2005. An adaptive solution to dynamic transport optimization. Presented at the Proceedings of the Fourth International Joint Conference on Autonomous agents and multi-agent systems, The Netherlands.

Fischer, K., Müller, J.P., Pischel, M., 1995. Cooperative Transportation Scheduling: an Application Domain for DAI. Applied Artificial Intelligence 10, 1–33.

Gendreau, M., Guertin, F., Potvin, J.-Y., SÈguin, R., 2006. Neighborhood search heuristics for a dynamic vehicle dispatching problem with pick-ups and deliveries. Transportation Research Part C: Emerging Technologies 14, 157–174.

Ghiani, G., Guerriero, F., Laporte, G., Musmanno, R., 2003. Real-time vehicle routing: Solution concepts, algorithms and parallel computing strategies. European Journal of Operational Research 151, 1–11.

Gutenschwager, K., Niklaus, C., Vo, S., 2004. Dispatching of an Electric Monorail System: Applying Metaheuristics to an Online Pickup and Delivery Problem. Transportation science 38, 434–446.

Gompel, J.V., Tuts, B., Claes, R., Torres, M.C., Holvoet, T., 2010. MAS-DiscoSim for PDP:a Testbed for Multi-Agent Solutions to PDPs. In Proceedings of the 9th International Conference on Autonomous Agents and Multiagent Systems: volume 1-Volume 1 (pp. 1639–1640). International Foundation for Autonomous Agents and Multiagent Systems.

Holvoet, T., Valckenaers, P., 2007. Exploiting the environment for coordinating agent intentions. Presented at the Proceedings of the Third International Conference on Environments for multi-agent systems III, Hakodate, Japan.

Holvoet, T., Weyns, D., Valckenaers, P., 2009. Patterns of Delegate MAS. Third IEEE International Conference on Self-Adaptive and Self-Organizing Systems. California, USA, San Francisco, pp. 1–9.

Jennings, N.R., 2000. On agent-based software engineering. Artificial intelligence 117, 277–296.

Maes, P., 1990. Situated agents can have goals. Robotics and Autonomous Systems 6.1, 49–70.

Mamei, M., Zambonelli, F., Leonardi, L., 2004. Co-fields: a physically inspired approach to motion coordination. Pervasive Computing, IEEE 3, 52–61.

Mes, M., Heijden, M. van der, Harten, A. van, 2007. Comparison of agent-based scheduling to look-ahead heuristics for real-time transportation problems. European Journal of Operational Research 181, 59–75.

Oluyomi, A., Karunasekera, S., Sterling, L., 2007. A comprehensive view of agent-oriented patterns. Autonomous Agents and Multi-Agent Systems 15, 337–377.

Parragh, S., Doerner, K., Hartl, R., 2008. A survey on pickup and delivery problems. Journal of Betriebswirtschaft 58, 81–117.

Personal rapid transportation system. http://en.wikipe dia .org/wiki/Personal_rapid_transit. (Accessed 25.11.2014)

Rao, A.S., Georgeff, M.P., 1995. BDI-agents: from theory to practice, in Proceedings of the First International Conference on Multi-Agent Systems.

Rodrigue, J.-P., Comtois, C., Slack, B., 2006. The geography of transport systems. New York, London, Routledge.

Shaw, M., Clements, P., 2006. The golden age of software architecture. IEEE Software 23, 31–39.

Tuts, B., 2010. Applying self-adaptation techniques to Delegate MAS. Master Thesis, Computer Science, K.U. Leuven, Leuven.

Wolf, T.D., Holvoet, T., 2007. Design patterns for decentralized coordination in self-organizing emergent systems in Proceedings of the Fourth International Conference on Engineering self-organizing systems. Hakodate, Japan, 28–49.

Wooldridge, M.J., 2002. An introduction to multi-agent systems, xviii, 348.

Xu, Y., Scerri, P., Yu, B., Okamoto, S., Lewis, M., Sycara, K., 2005. An integrated token-based algorithm for scalable coordination presented at the Proceedings of the Fourth International Joint Conference on Autonomous agents and multi-agent systems, The Netherlands.

CHAPTER 6

Studying the Impact of the Organizational Structure on Airline Operations Control

Nuno Machado*, António J.M. Castro**, Eugénio Oliveira†

*MIEIC, DEI, Faculdade de Engenharia da Universidade do Porto, Porto, Portugal; **LIACC, Faculdade de Engenharia da Universidade do Porto, Porto, Portugal; †LIACC, DEI, Faculdade de Engenharia da Universidade do Porto, Porto, Portugal*

6.1 Introduction

An organizational structure might be regarded as a set of entities collectively collaborating and contributing toward one common goal. The employees working on an assembly line or a rescue team are examples of organizational structures. Nowadays, with the increasing complexity of goods and services, and competing in a globalized world, organizations require tuned work systems, involving human capital interwoven with the latest technological innovations.

Evolving an established organizational structure is often daunting when it is behind the core mission of a business or when it operates uninterruptedly. In these cases, software simulations are an invaluable tool to explore new work practices, information flows or even decision-making processes. Modeling and simulating complete or small portions of critical workflows make it feasible to collect a set of metrics as well as introducing organizational transformations. Brought together, these factors allow for organizational performance assessment and evolution.

The work presented in this chapter is founded on such observations and aimed at proposing improvements to the operational control within a real airline company. To accomplish such an aim, we had to use a real airline company as case study. TAP, the major Portuguese air carrier, agreed to participate in such a project and provided useful information and data. As any simulation-based research, this study involved three main stages that will be discussed in the following sections. First, we had to unveil the entities involved in airline operations such as facilities, supporting systems, human collaborators, and their main activities. Next, we used Brahms, a multi-agent system featuring the BDI model and its own agent-oriented programming language to model and simulate the airline's empirical concepts. Finally, we collected a set of metrics, introduced organizational structure modifications, and established a quantitative comparison among the latter.

Copyright © 2015 Zhejiang University Press Co., Ltd. Published by Elsevier Inc. All rights reserved.

6.2 Background

First and foremost, we tried to get some background information or discover targeted literature about other initiatives regarding airline operations control simulation but to no avail.

Following this, to the best of our knowledge we were the first to simulate the Airline Operational Control Center (AOCC) organizational structure in order to study its impact on airline disruption handling. Because of that it is difficult to compare our approach with others. Nevertheless, in this section we would like to provide some background regarding work systems modeling and simulation, and also about AOCC organization and some work related to disruption management.

6.2.1 Work Systems Modeling and Simulation

A work system involves people engaging in activities over time. Human participants might not just interact with each other, but also with machines, tools, documents, and other artifacts (Calvin and Pava, 1984).

The activities performed often produce goods, services, or data. There are two different approaches when it comes to designing or improving systems: (1) machine-centered and (2) human-centered (Sierhuis and Clancey, 2002). The former is usually accomplished through a business process reengineering approach (Mayer et al., 1998) based on business process flow analysis focused on work products. The latter also takes into account how the people in the organization actually prefer to work (Greenbaum and Kyng, 1991). Unlike the machine-centered approach, which neglects human communication, collaboration, workspaces, problem solving and learning; the human-centered approach analyzes human activities, work processes or tasks, comprehensively and chronologically throughout the day (Clancey, 2002).

The human-centered work system design approach is based on modeling and simulating work practices: what people actually do, rather than their outcomes. This way, it is possible to understand the effects of human behavior in different places and times, details often omitted in a product-oriented task analysis. In the end, besides the traditional system workflow, the human-centered approach might also propose some work system transformations, including different tools, resources, locations, or scheduling.

Aiming at using a human-centered approach to model and simulate our organizational structures, Brahms (Sierhuis, 2001) was adopted as modeling and simulating tool. It follows a holistic approach to systems modeling. By developing formal models of people's behavior at the activity level, it is possible to determine the impact of these actions on the whole system.

Besides its own agent-oriented programming language, Brahms contains some predefined model components that make it straightforward to implement reality concepts:

- *Agent/groups*: to model the human collaborator;
- *Objects*: for the computerized systems;

- *Geographies*: used to indicate the location of facilities;
- *Activity*: to express the agent behavior;
- *Timing/workframes*: used to model activity duration.

Brahms does not provide real-time visual feedback of a running simulation. Therefore, this deficiency had to be addressed through the development of a visualization of the simulated airline.

6.2.2 Airline Operational Control Center Organization

The main role of the AOCC is to monitor the conformance of fight activity according to the previously defined schedule. The occurrence of some unexpected events might prevent operations taking place as planned, such as aircraft malfunction, crew delays, crewmembers absence, and so forth.

Following this, the AOCC is a human-decision system composed by teams of experts specialized in solving the described problems. Teams act under the supervision of an operational control manager and their goal is to restore airline operations in the minimum frame and at a minimum cost.

According to Castro (Castro, 2008), there are three main AOCC organizations:

- *Decision center*: the aircraft controllers share the same physical space. The other roles or support functions (crew control, maintenance service, etc.) are in a different physical space. In this type of Collective Organization all roles need to cooperate to achieve the common goal;
- *Integrated center*: all roles share the same physical space and are hierarchically dependent on a supervisor. For small companies we have a Simple Hierarchy Organization. For bigger companies we have a Multidimensional Hierarchy Organization. Figure 6.1 shows an example of this kind of AOCC organization;
- *Hub control center*: most of the roles are physically separated at the airports where the airline companies operate an hub. In this case, if the aircraft controller role stays physically outside the hub we have an organization called Decision Center with a hub. If both the aircraft controller and crew controller roles are physically outside the hub we have an organization called Integrated Center with a hub. The main advantage of this kind of organization is to have the roles that are related to airport operations (customer service, catering, cleaning, passengers transfer, etc.) physically closer to the operation.

As mentioned, Figure 6.1 shows the traditional Integrated Operational Control Center. As previously stated, the AOCC is composed of groups of workers, each one with its own responsibilities. They must report their activity to a Supervisor, translating a two-level hierarchical system. Figure 6.1 also represents the activity time-window of the AOCC, it starts 72–24 h before the day of operations and ends 12–24 h after.

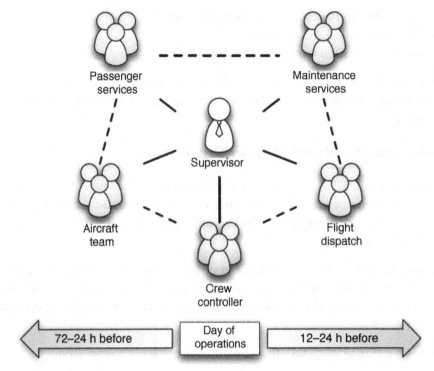

Figure 6.1: Integrated airline operational control center.
(*Source*: *Adapted from Castro and Oliveira (2010)*)

The roles most common in an AOCC are, according to Kohl and Karisch (2004) and Castro (2008):

- *Flight dispatch*: prepares the flight plans and requests new flight slots to the Air Traffic Control (ATC) entities (FAA in North America and EURO-CONTROL in Europe, for example);
- *Aircraft control*: manages the resource aircraft. It is the central coordination role in the operational control;
- *Crew control*: manages the resource crew. Monitors the crew check-in and checkout, updates and changes the crew roster according to the arisen disruptions;
- *Maintenance services*: are responsible for the unplanned services and for the short-term maintenance scheduling. Changes in aircraft rotations may impact the short-term maintenance (maintenance cannot be done at all stations);
- *Passenger services*: decisions taken by the AOCC will have an impact on passengers. The responsibility of this role is to consider and minimize the impact of the decisions on passengers. Typically this role is performed by the airports and for bigger companies is part of the HCC organization.

6.2.3 Disruption Management

Disruption Management (Kohl and Karisch, 2004), also known as Operations Recovery, is the process carried out by the AOCC when an unexpected problem prevents a flight operating as planned.

The first overview of the state-of-the-practice in operations control centers on the aftermath of irregular operations and was provided by Clarke (1998). In his study, besides an extensive review of the subject, he proposes a decision framework that addresses how airlines can reassign aircraft to scheduled flights after a disruptive situation.

Currently, the most thorough analysis of the discipline is presented by Kohl et al. (2007) where their conclusions are supported by the DESCARTES project, a large-scale airline disruption management research and development study supported by the European Union.

Other authors propose more general perspectives regarding disruption management. Yu and Qi (2004) analyze airline disruption management from different angles: crew and aircraft recovery; and applied to other fields as well: machine scheduling and supply chain coordination. Given the large scope of their work, airline operations recovery is not particularly detailed.

On the other hand, Ball et al. (2007) give insight into the infrastructure and constraints of airline operations, as well as the air traffic flow management methods and actions. Simulation and optimization models for aircraft, crew and passenger recovery are also discussed. Furthermore, the authors give an excellent survey of the airline schedule robustness as a proactive alternative to recovery, including model descriptions and a literature review.

From the mentioned studies, it is clearly a tendency to consider the disruption management problem as twofold: aircraft recovery and crew recovery. For each type of recovery several solution approaches were proposed on the basis of different methodologies.

An in-depth and comprehensive review of the most relevant studies and methodologies used in disruption management is presented by Clausen et al. (2010). They not only explain the most traditional approaches, such as Connection, Time Line, and Time Band Networks, based on the scheduled aircraft and crew rosters but also mention newer and innovative research studies.

Although the vast majority of the publications use integer programming solution methods to solve the aircraft recovery problem, the most recent works apply some metaheuristics to the problem, such as described by Andersson (2006) and Liu et al. (2006).

Moving to crew recovery, the majority of publications formulate the crew recovery problem under the assumption that the flight schedule is recovered before the crew rescheduling decisions are made, thereby following the hierarchical structure of the disruption recovery in practice. These publications include Wei et al. (1997), Guo (2005), and Nissen and Haase (2006).

For instance, from the list of authors presented in the last paragraph, Wei et al. (1997) model the crew pairing repair problem as an integer multicommodity network flow problem on a Connection Network. The challenge is to repair the pairings that are broken and the objective is to return the entire system to the original schedule as soon as possible while minimizing the operational cost.

Something interesting about Nissen and Haase (2006) research is that it is based on European reality. They propose a duty-based formulation for the crew recovery problem, which is especially well suited for solving crew disruption for European airlines, as these, contrary to the North American airlines, employ fixed monthly crew rates, which should be taken into consideration when solving a crew disruption.

Finally, (Castro and Oliveira (2010) pioneer an approach that not only accounts for the aircraft and crew perspectives but also considers passengers. An implementation of an intelligent and distributed multi-agent system (MAS) represents the operations control center of an airline. MAS includes a crew recovery agent, an aircraft recovery agent and a passenger recovery agent. They use concepts of direct and qualitative cost to determine solutions for the disruption problem.

6.3 Empirical Airline Operations

The airline operations start way before the actual flight day as they require the scheduling of flights in advance. Then several stages emerge such as the revenue management, aircraft and crew rosters, and so on (Antonio, 2010). This is usually known as the Airline Scheduling Problem (Grosche, 2009).

When the day of operations arrives, unexpected events may prevent flights to depart as planned and the airline specialists must address those situations. This is known as the disruption management problem.

Our study is about organizational structures of the AOCC in the context of the day of operations, not to the disruption management algorithms and/or processes that are used to solve the disruptions. For that, we need to know the workflows before and after that stage of the disruption, that is, which are the unexpected events, who detects such events, how the airline specialists know about them, and who is notified of putative solutions.

In order to simulate such a scenario we needed to know the entities involved in airline operations. Figure 6.2 clearly depicts those entities and their geo-location. Squares represent facilities and ellipses computerized systems. Table 6.1 describes each of the entities' labels.

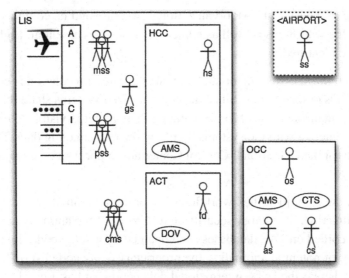

Figure 6.2: Current TAP organizational structure

Table 6.1: Organizational structure concepts.

Facilities	
ACT	Crew Terminal
AP	Aircraft Parking
CI	Passenger Check-In
HCC	Hub Control Center
LIS	Lisbon Airport
OCC	Operational Control Center
Computerized systems	
AMS	Aircraft Movement System
CTS	Crew Tracking System
DOV	Flight Operations Portal
Human collaborators	
as	Aircraft Specialist
cms	Crew Members
cs	Crew Specialist
fd	Flight Dispatcher
gs	Ground Supervisor
hs	HCC Supervisor
mss	Maintenance Services
os	OCC Supervisor
pss	Passenger Services
ss	Station Supervisor

With a big picture of the current TAP organizational structure and its components, an in-depth understanding of the workflows as well as related activities was essential. Eight workflows and activities were identified.

Concerning the activities, across the organizational structure, information is conveyed by means of VHF radios or telephones. Since the computerized systems share the same network, information is instantaneously synchronized among them and it is visible to each other. Human collaborators interact with the systems by filling forms or reading data. The specialists carry out decisions at the Operational Control Center and supervisors are required to approve those decisions.

Triggering any of the eight identified workflows is a preflight anomaly, for example, lack of fuel, aircraft malfunction, mandatory security, and so forth. If the anomaly causes a departure delay, then it is recorded on TAP databases accompanied by a delay code. Just to provide an idea of the number of potential anomalies, the proprietary delay code list of TAP has more than 200 entries, whereas the IATA international delay code list has around 80 anomaly types. An airline operator or system usually detects each anomaly, thus inquiries were made in order to classify each delay code according to concept.

In order to illustrate an operational workflow, an example follows. Imagine that 15 min before departure a ULD (Unit Load Device), inadvertently hits an aircraft during cargo loading. Assuming that this kind of anomaly has a delay code of 100 and TAP had classified such a code as being detected by the Ground Supervisor then, at this point, the Ground Supervisor is the only agent knowing about the problem. The deciding agents in the TAP organizational structure are the Aircraft and Crew Specialists, located at the OCC. They must be aware of the problem in order to find the best solution, for example, replace the aircraft, delay the flight, and so forth. Figure 6.3 illustrates the workflow behind the resolution of an aircraft anomaly detected by the Ground Supervisor.

In order to alert the Specialists, the Ground Supervisor first uses the VHF radio to communicate the problem to the HCC Supervisor. Next the latter fills a form to enter into the Aircraft Movement System and the information is propagated instantaneously to the OCC. There, the Aircraft Specialist is hopefully paying attention to the screen and becomes aware of the problem. He seeks a reason for the problem and after reaching a conclusion inputs it into the AMS, being replicated to the CTS. Now it is the turn of the Crew Specialist. Mandating or not some crew assignment changes, the Crew Specialist is required to evaluate, take action and confirm the solution suggested by the Aircraft Specialist through the CTS terminal. His input will be readily synchronized, once again, with the Aircraft Movement System, making it available to both OCC Supervisor and HCC Supervisor. As the main character in the Operational Control Center, the OCC Supervisor is required to ratify the decisions proposed by the Specialists, whereas the Hub Control Center Supervisor uses the VHF radio again to communicate changes to the Ground Supervisor.

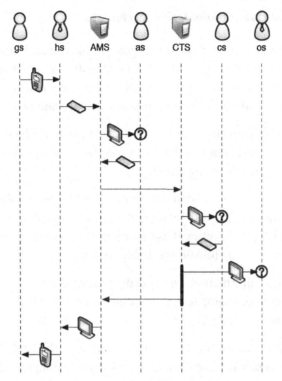

Figure 6.3: Workflow triggered by an AC anomaly detected by Ground Supervisor

All the activities above require time to perform. TAP was questioned about the duration of such activities and; although a definite answer was impossible, it provided minimum and maximum time intervals for each activity. At this point we understood that communications by phone take, on average, more time than VHF radio transmissions as they are usually concerned with more complex anomalies.

6.4 Organizational Structure Performance Assessment

This section intends to suppress the lack of information regarding the organizational structure simulation by presenting the main features of Brahms, the modeling and simulating tool that introduces a new human-centered computing paradigm. In order to accomplish this, we will first draw a big picture of the simulation system as a whole justifying the use of certain technologies and pointing out additional contributions to the communities behind those technologies. Further, we will explain how Brahms greatly improved the experience of modeling the workflows at TAP. Finally, a section will also be dedicated to expose some aspects of a visualization module developed to allow a better understanding of the concepts being simulated.

6.4.1 Background and Overall Simulation Architecture

As referred to during the introduction, the main goal of this research study was to simulate the operational control of a real airline company. Obviously, simply creating a model of such reality and mimic its intrinsic features would be of arguable interest so we aimed thereafter at proposing changes that would lead to more efficient activities and workflows.

The empirical observations listed in Section 6.3 made us aware of the reality at TAP, our case study airline company. We soon noticed that we would be treating a case that falls into the popular business-process reengineering paradigm.

Following this, we had to adopt a simulation tool that would simplify the modeling of the concepts related to our airline company while at the same time featuring some business process reengineering capabilities. Meeting these requirements was Brahms (Sierhuis, 2001) the Business Redesign Agent-Based Holistic Modeling System.

Although some theoretical information was already presented about Brahms in the Background, it is worth pointing out some technical information about this system for the purpose of clarifying certain options or side activities carried out alongside this study.

First and foremost, Brahms[1] is a Multi-Agent System featuring the BDI (Beliefs, Desires, Intentions) architecture. Whereas these characteristics are not enough to distinguish it from many other simulation engines, the Brahms Team at NASA Ames Research Center is currently developing Brahms in collaboration with the Carnegie Mellon University, and it has been successfully used in NASA's Mission Control to automate human tasks for the International Space Station. Its source code is proprietary but NASA freely distributes it for research purposes only.

At this point, we simply thought that if Brahms were enough for NASA it would certainly suit our needs. After further inspecting the features provided by Brahms, we noticed that it was a much more advanced tool than other Multi-Agent Systems that we knew about. It sports its own agent-oriented programming language, adds up some human-centered computing concepts and has its own production rules system.

The characteristics above ought to require an additional effort in implementing it in our airline company, but we decided to take the challenge. Another feature lacking in Brahms is the ability to visualize the concepts being simulated. Although this functionality was not required to get a quantitative comparison of different organizational structures, it was regarded as an educational and clarifying way of understanding the operations carried out in an airline company.

Being mostly characterized as an academic and scientific tool, Brahms lacks a wide user community, where one may get models, code examples and ask for help. In spite of that, it is thoroughly documented and their creators lead a discussion group to assist early-adopters.

---

[1] Software model that implements the main aspects of Michael Bratman's theory of human practical reasoning.

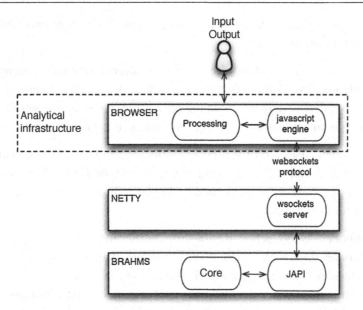

Figure 6.4: Overall simulation architecture and components

As Brahms runs in a closed virtual machine, the first contact with the simulator community was intended to inquire about the possibilities of developing a visualization of a running simulation. Although the primary approach would be to interpret a set of output files postsimulation, Brahms features a Java API (JAPI) allowing for environment expansion and control.

Interacting with JAPI would be roughly the same as interacting with a Java application. Therefore, we decided that our visualization would be built-in in the analytical infrastructure and use the latest advancements in browser technology.

Figure 6.4 depicts the components and architecture beneath the simulation portion of our study. Although this section is not meant to be too technical, other aspects of Figure 6.4 require further inspection. Starting from the beginning, the human user has the ability to interact with the analytical infrastructure through a browser. Given the set of technologies used, it is important to emphasize that at the time of writing, the only browser that supported our infrastructure was Google Chrome. Nevertheless, with fast technological evolution, it is likely in the near future other major browsers start to implement the technical innovations employed by our simulations.

Moving down in Figure 6.4, the browser portion of the simulation architecture contains the visualization module. It is mainly composed of two interwoven parts: the "javascript engine" and "processing". The former handles all the communications with the "websockets server", discussed soon, decoding the incoming messages and controlling "processing" animations.

Processing is widely used in the scientific and academic field, given its ability to create powerful representations of large sets of data. Although the original processing is based and to be used with the Java programming language, considering our browser requirements, we had to use a javascript port of the language.

The BROWSER component also allows for simulation control, which is, starting, pausing, and stopping the simulation. The visualization module will be described in Section 4.3, so at this point it just matters understanding we are in the presence of a distributed infrastructure where messages come in, go out and an animation of the simulated theatre is displayed in-between.

The "websockets server" uses NETTY, a Java nonblocking I/O socket framework, to implement the recently introduced *websockets protocol* as part of the HTML5 specification. The use of such technology is solely implemented on Google Chrome, thus the reason our simulations only work with this browser.

The NETTY component is too technical to deserve further inspection. As any server, it establishes TCP connections with remote clients and then exchanges messages with the Java API of Brahms. Besides a message gateway, it shares some similarities with the "javascript engine" as it decodes and encodes the messages exchanged with the simulation core.

The next section will be solely concerned with the next component, BRAHMS. As we referred to, the selected simulation engine to implement our organizational structure features an agent-environment, "core" and a Java API, "JAPI". The former is more concerned with agent, geography, activities and other real entities modeling; the latter is more technical, being used for handling Java objects or other services.

6.4.2 The Simulation Module

This section is focused on the BRAHMS component of the simulation architecture (Figure 6.4). We will start by describing the "core", that is how we used Brahms formalisms and programming language to implement the empirical observation exposed in Section 6.3. As stated, the "JAPI" side is more technical, therefore we will not delve deeply into it, solely pointing out its use as a means to solve some Brahms shortcomings.

Brahms supplies a number of human-centered structural formalisms to help modeling real entities. Thus, one of the first steps in modeling a scenario with Brahms is to make a correspondence between real and artificial concepts. Table 6.2 intends to clarify our approach concerning such mapping.

Besides presenting a number of associations, Table 6.2 also hopes to illustrate the expressiveness offered by the Brahms modeling language. Although comprehensive, it just contains a subset of Brahms concepts.

The first column respects the reality, the next two contain the name of virtual entities implemented in our simulations. Starting by "Facilities and Locations", Brahms is very complete in what concerns geography modeling. There are areas and areadefs and we may not have the

Table 6.2: Concept mapping between reality and Brahms formalisms.

Reality	Brahms	
Facilities and locations	*area*	*areadef*
ACT	LisbonAirportACT	ACT (Builing)
AP	LisbonAirportAP	AP (BaseAreaDef)
CI	LisbonAirportCI	CI (Building)
HCC	LisbonAirportHCC	HCC (Building)
LIS	LisbonAirport	Airport (BaseAreaDef)
OCC	TapOCC	OCC (Building)
Computerized systems	*object*	*location*
AMS	HCC_AMS	LisbonAirportHCC
	OCC_AMS	TapOCC
CTS	OCC_CTS	TapOCC
DOV	ACT_DOV	LisbonAirportACT
(AS)	Lisbon_AS	(World)
Human collaborators	*agents*	*groups*
as	AircraftSpecialist	OCCSpecialists
cms	CrewMember	CrewMembers
cs	CrewSpecialist	OCCSpecialists
fd	FlightDispatcher	TriggeringAgents
gs	GroundSupervisor	GroundPersonnel, TriggeringAgents
hs	HccSupervisor	ApprovingAgents
mss	MaintenanceMan	GroundPersonnel, TriggeringAgents
os	OccSupervisor	ApprovingAgents
pss	PassengerMan	TriggeringAgents
ss	StationSupervisor	TriggeringAgents13

former without the latter. That is, first the area must be defined, we must specify what it is then we may name it. Looking at the "ACT" example, we first had to create a generic "ACT" extending the Building definition shipped with Brahms and then we were able to define "LisbonAirportACT" as an instance of it. In the case of "AP", Aircraft Parking, as it is not a Building, we had to choose the BaseAreaDef as parent.

Our naming conventions reveal another Brahms feature not foreseeable in the table, the "part of" construct. Listing 6.1 shows an excerpt of the "part of" and path formalisms.

Listing 6.1 Excerpt of Brahms area and path definitions

```
area   LisbonAirportAP   instanceof   AP   partof   LisbonAirport   {

}

path   LisAP_to_from_LisHCC   {

    area 1 :   LisbonAirportAP ;

    area 2 :   LisbonAirportHCC ;

    distance :    600 ;

}
```

As is clear, besides specifying what the area is, we may also specify the relation between areas (part of). Figure 6.2, shows the Aircraft Parking inside the Lisbon Airport and Brahms allows such modeling. Another very convenient feature is the path. It defines a relationship between two areas not in terms of composition but geographical dispersion; again, something very handy to set the distances (in s) between buildings or areas. The distance, on average, between the Aircraft Parking and the Hub Control Center is 10 min, so we must specify it as 600 s. As we will see later, if our agents use a vehicle, and thus only spend 2 min, the 600 s time may be overwritten in the move activity.

Moving on to the next portion of Table 6.2, it is about computerized systems. To model inanimate things, Brahms offers the concept of object. Actually, some real-world objects might be modeled as agents because, although they are physically inanimate, they may be used to reason over facts and therefore help humans take decisions, for example, a computer. The notion of agent in Brahms is a little narrower than other multi-agent systems because objects might also react and reason as agents. Apart from the naming convention used for systems, which is irrelevant, another key property is its location. In Brahms every object and agent might be given a location. Again, according to Figure 6.2, the airline systems were distributed across different locations.

Concerning the human collaborators, as stated above they were modeled as agents. As a multi-agent system, this is no surprise. The innovative factor in Brahms is the existence of groups. When implementing an agent, one may use the "memberof" keyword to set its group membership. For instance, in Table 6.2, the "AircraftSpecialist" and the "CrewSpecialist" are members of the same group, the "OCCSpecialists". This is a very powerful feature in Brahms because when we need to implement activities to be performed by the agents, we just need to implement them at the group level, then the activities are automatically inherited.

Before introducing Brahms activities, we may not skip the "Lisbon AS" object. The "AS" systems stands for Airport Screen and during our interviews with airline personnel nobody noticed its existence, thus the reason for appearing between parentheses. It is here to illustrate a simple case where a modeler needed to use a workaround to simplify or make it computationally feasible to mimic reality.

Recalling our empirical scenario, when some agents detected anomalies they would trigger workflows. In Table 6.2 they appear as members of the "TriggeringAgents" group. In reality, they perceive anomalies in the course of their activities: verifying an aircraft, checking-in passengers, loading cargo, and so on. But in our simulation we only had files with those anomalies. The closer approach would be to have those agents all reading the file and checking if they were responsible to trigger the next anomaly. Although it would be correct to do it that way, it would not be wiser because it would put too much strain on IO operations to read the same file (or check the same list), over and over again.

Following this, we created the Airport Screen system that roughly mimics those screens found at the airports with the next departures or flight delays. It reads the file and "tells" the agents about upcoming anomalies. Now that hopefully the notion of auxiliary object was explained how about other topics worth discussion: how does the Airport Screen tell the other agents?

Answering this question definitely proves that Brahms is a fairly different multi-agent system founded on a totally new paradigm. As any other programming language, the Brahms agent-oriented programming language also supports the primitive types, such as integers, characters, etc. and the map collection. Unfortunately, it does not support lists, a major flaw that had to be overcome through the use of JAPI, a workaround explained soon. Although Brahms supports those data types, agents, and objects are unable to directly handle them. At this point is very important to underline that Brahms is human-centered and human beings do not act or reason upon integers, they do that according to facts or beliefs. This is where the BDI software model enters and somehow distinguishes Brahms from the majority of multi-agent systems.

Now that facts and beliefs were introduced, when our Airport Screen detects an anomaly it creates a fact or belief. As it is located in the "World", a Brahms abstraction to everywhere, all the agents or objects perceive such fact/belief. It is up to the modeler to implement the activities they must perform, if any, when they detect the fact.

To better illustrate the human-centered paradigm of Brahms, Listing 6.2 contains a purportedly oversimplified code excerpt. It shows a routine that every minute checks the flight list using a Java object (more about this later) and concludes a fact, triggerConcept, represented by the conceptid string. In the TriggeringAgents when the triggerConcept fact matches the conceptCode the agent does something.

Up until now we have presented an overview about how we modeled facilities, systems, and collaborators. The next step is to summarily expose the Brahms formalisms concerned with activities.

Listing 6.2 The Brahms human-centered paradigm

```
//   in   Lisbon_AS ...
repeat :   true ;
when ( knownval ( current . currMinute  > current . cfMinute ))
do  {
string   conceptid  = as. checkFlights ();
conclude (( current . triggerConcept  =  conceptid ), bc :0 ,  fc :100);
}
//  in  TriggeringAgents  group ...
when ( knownval ( Lisbon_AS . triggerConcept  =  current . conceptCode ))
do  {
...
}
//  in  GroundSupervisor  agent ...
initial_facts :
( current . conceptCode  =  "gs ");
```

We identified six main activities: communicate (by radio and phone), data write, data read, reasoning, and approve. To these six, let us add a new one that will be used in our future organizational structure proposals: move between locations.

Brahms supports multiple activities, and one of them is the Java activity. The Java activity will not be discussed in detail but it is worth mentioning because it might be regarded as executing a Java method, with inputs and multiple return values. As such, it virtually allows Brahms to achieve anything possible with Java. For instance, the activity of checking flights on Listing 6.2, which required us to read from a file, could have been implemented using a Java activity.

Other types of activities, more relevant to our study were the *communicate* and the *move* activities. Concerning the former, what we knew was that certain airline operators, in the presence of an anomaly, would pick up the radio or phone and communicate such a fact to a supervisor. We also knew that such activity would consume an indefinite amount of time.

As in other systems, there are always a number of ways to implement the same scenario and our simulation was no exception. There would be several ways of communicating a disrupted flight but in our case we opted for the flight number. Ideally, it would have been better to pass a Java object because, as we will see soon, our flights were implemented as such. The problem is that agents in Brahms, as human counterparts, are solely capable of transmitting facts or beliefs, usually represented through primitive data types.

Listing 6.3 intends to show how easily Brahms makes the transmission of facts and beliefs across agents. The excerpt presented is, again, part of the TriggeringAgents group, therefore it will be inherited by multiple agents, each one with its own recipient. To surpass this issue the communicate activity shown resembles a function where the "with" field is variable.

Still on Listing 6.3 the "about" field indicates the fact or belief to be sent, in this case the disruptedFlightNumber. Once in possession of the fact or belief, the recipientAgent may act or reason upon it. Last but not least, there is the activity duration. By asserting the "random" property as true, we want Brahms to pick a value between the "min duration" and the "max duration".

Listing 6.3. The Brahms communicate activity

```
communicate   reportDisruptedFlightByPhone ( BaseGroup   recipientAgent ) {
with :    recipientAgent ;
about :    send ( current . disruptedFlightNumber );
random :    true ;
min_duration :    240;
max_duration :    480;
}
```

The way Brahms handles activity timing was of utmost importance for our study. The other activity types benefit from the same random approach and therefore the previously seen "distance" in geography paths (see Listing 6.1), may be overwritten using a move activity.

The move activity is not much different from the communicate activity, instead of a "with" and "about" properties, it has a "location" property telling the agent where to go next. The motion takes a certain amount of time that might be random, as in Listing 6.3 or static, asserting "random" as false and providing a "max duration".

Before moving to the JAPI component of the simulation architecture (refer back to Figure 6.4), a brief word goes to Brahms classes. Along with areas, objects, groups and agents, Brahms also supports classes. The problem is, these classes are not as powerful as the Java counterparts. Actually, they use the same Brahms agent-oriented programming language syntax, and the same human-centered paradigm. Therefore and simply put, classes are to objects as groups are to agents. We did not list the classes in Table 6.2 as there is solely one, the TriggeringObjects that works in a similar fashion to TriggeringAgents.

Up until now we described our approach by what concerns the modeling of the most visible concepts and activities using the Brahms proprietary agent-oriented language. Although we recognized how expressive, distinct, innovative, and somehow powerful it is, we must also underline its shallow learning curve and, as we will see further, the lack of some widely used data types and support functions.

As we stated previously, Brahms supports several primitive data types and maps. Unfortunately, lists are not available and they are one of the key data structures to store our flights. Even the flight object, which is composed of several attributes, such as scheduled departure date or flight number, would be much better abstracted by means of plain Java objects.

To address such issues Brahms provides two options. The latest alpha version allows for direct Java objects manipulation. Older versions support already mentioned Java activities. In one case or the other, there are some conventions one should respect but in the end it is roughly like calling static Java methods.

Without getting into much detail, in the implementation of our simulations, the Brahms JAPI was used in several scenarios. First and foremost, to store within ArrayLists our flights and delays objects. Second, to perform file input and output, operations not supported at the Brahms level. Third, to implement the Specialists reasoning activities. Last to stream the ongoing events to the "websockets server", see Figure 6.4.

To conclude, it is worth emphasizing that this section did not aim at thoroughly describing the implementation of our simulation using Brahms. That would require a technical manual as long as this report. The intention here was to present the human-centered nature of Brahms and how that paradigm fits the reality being modeled.

6.4.3 The Visualization Module

As a side goal, the visualization module was not required to produce the answers to the main goals of this research study. It simply receives some messages from the Brahms component, such as which activity is being carried out by which agent, and displays an animation of the simulated theatre. Therefore, the main purpose of the visualization was to provide an educational tool to allow people to learn how the airline operations management works, as well as to better understand the proposed organizational changes.

As was explained in Section 4.1, the visualization module uses Processing.js to render images and animations on the recently introduced HTML5 canvas. It is tightly connected to the javascript that decodes the messages coming from the simulation.

The visualization is composed of two distinct areas, the operational area, very similar to Figure 6.2, and an airport screen. The former is where the main action takes place, through arrows we may observe the current workflow state. The latter provides visual hints about flight departures and state. The airport screen lists all the flights within a future time frame, if a flight suffers an anomaly it is depicted in a different color and a workflow is triggered in the operational area. Assuming it was the Ground Supervisor who detected the anomaly, his next activity is to notify the HCC Supervisor, and then an arrow is displayed between him and the HCC Supervisor with a visual indicator of a radio communication.

Without being too technical, the underlying architecture of the visualization had to closely implement the concepts introduced by Brahms. This means we had to implement classes to represent the agents, the objects, the area definitions, and so on. Although requiring an additional effort, such an approach also allows for a flexible display.

Given the need to represent several distinct organizational structures, the visualization had to be dependent on the simulation. At the beginning, a list of the Brahms concepts is passed to the visualization so they can be displayed. Other features are also present such as onMouseOver actions that return further information about the concepts and so on.

To conclude, a final word goes to the amount of Processing code required to implement solely one action or even the Brahms concept. We must keep in mind that behind a Brahms concept abstraction there is a complex and large code base, therefore the need to execute the Brahms environment in a virtual machine. The problem is that we lack such constructions in Processing and we are required to implement them by hand. Following this, to implement every activity or concept is a lengthy process, the reason why the visualization will always be less expressive than the simulation itself.

6.5 Scenario and Experiments

This section aims at presenting the underlying aspects of simulation input, transformation, and output. It provides useful insights to understand the organizational results presented in the next section.

6.5.1 Simulation Input Data

As advertised, our simulations used real operational data from TAP. In the context of our research, a database service was purportedly implemented to collect pre- and postflight activity. The preoperational records included the scheduled flights, assigned aircraft, and assigned crewmembers. On the other hand, postoperational data exposed the flights that actually took off as well as aircraft and crew changes. We were also given a list with all the flights that suffered departure delays, the number of minutes, and the corresponding TAP and IATA delay codes.

Possessing such data allowed us to input the scheduled flights and treat the delays, recorded after operation, as anomalies occurring during flight handling, that is, an actual flight that suffered a delay caused by unexpected late passenger check-in, would be simulated as suffering a late passenger check-in anomaly.

It is worth emphasizing the utmost importance of using real data. In an organizational structure not all the business processes assume the same prevalence, for example, there are workflows that take place a higher number of times than others. Since we will use anomalies to trigger workflow execution, using random data would not respect the uneven distribution of processes, compromising the final results.

Our simulation was fed with the flights operated by TAP from the 15th to the 21st February of 2010, a whole seven days of activity. Although 7317 flights were scheduled to take place that week, due to data incompleteness, for example, missing databases fields, table referential deficiency, inconsistent data, we were only able to input 1801 flights, 389 of which suffered anomalies.

6.5.2 Operational Workflow Transformations

The major goal of our study was to assess distinct airline organizational structures. On the basis of the actual airline simulation, the control group, three organizational structures were incrementally changed and simulated. All the simulations were executed after the same operational scenario, comprising the scheduled flights and anomalies referred to in the previous section. When proposing organizational structure modifications we were cautious not to alter the inputs and outputs of the business process, that is, never change the triggering and deciding agents.

Our first proposal (1) suggested the removal of the HCC Supervisor. After analyzing the actual sequence diagrams, we observed that he usually plays a part as information distributor and only assumes a supervising position when facing anomalies related to Passenger Services. Removing the HCC Supervisor required three major changes in four (out of the eight) workflows. The Ground Personnel were now required to go to the Hub Control Center to input data into the AMS; OCC Supervisor accumulated the role of notifying Ground Personnel about OCC Specialists decisions; and the Passenger Services started to report anomalies to the OCC Supervisor via phone.

Proposal 2 departed from proposal 1 and aimed at avoiding the Ground Personnel to go to the Hub Control Center in order to reach the Aircraft Movement System. This way, we suggested adding mobility support to the existing AMS, making it manageable through a wireless smartphone or laptop. Conscious of certain security implications, we decided that at this stage access would be solely granted to Ground Personnel. All the remaining operators kept interacting with AMS the same way as they did previously.

In our last proposal 3, we removed the usage restrictions on the AMS found in proposal 2 and started to think of it as a web-based system accessible from everywhere. At this stage, the Flight Dispatcher and the Station Supervisor were now able to input and read data from the AMS, no matter their location.

6.5.3 Metrics

Two metrics were used to assess organizational structure performance: overall disruption handling time and average collaborator stress. Although they are both based on the activity duration, they measure different concepts. Overall disruption handling time is the sum of the time consumed by all the workflows, that is, when an anomaly disrupts a flight it also triggers a workflow composed of several activities, which durations will be summed up until a solution for the anomaly is found. Concerning collaborator stress, it is a metric associated with each collaborator and thus requires a statistical aggregation to be used, for example, the average. It measures the number of hours spent by a collaborator in the course of a simulation.

There are activities that contribute only once to the overall disruption-handling time but several times to the collaborator stress. For instance, a phone call duration is added once to the former, but contributes twice to the overall stress, once per agent involved in the communication.

6.6 Results and Conclusion

Considering the scenarios depicted in the previous section, Figure 6.5 presents the comparison across proposals of the overall disruption handling time (left) and the average operator stress (right). The measurements are carried out in hours.

As expected, the metrics in analysis show a certain correlation, even though the collaborator stress is more affected by organizational structure transformations. The proposal that performed better was the third, achieving an improvement of 15% in the overall disruption handling and 21% for collaborator stress.

Figure 6.6 compares stress across collaborators and the proposal (chart column labels described in Table 6.1).

As one may observe, the OCC specialists ("as" and "cs") stress remained the same across all proposals since they are deciding agents at the center of the airline workflows. In the first

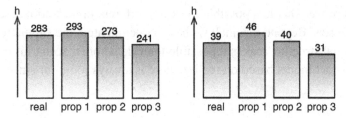

Figure 6.5: Overall disruption handling time (left) and average operator stress (right) across proposals

proposal "hs" was subtracted and "gs", "mss", and "os" suffered the highest impact. In the second, the wireless Intranet capabilities introduced in AMS, allowed the stress results to get back to the real values, except for "os". The last proposal, transform the AMS into an Internet-based system caused the highest general impact on stress.

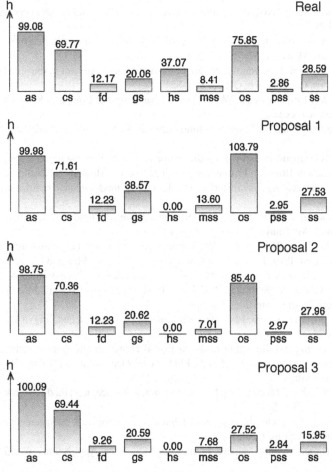

Figure 6.6: Comparison between collaborator stress and proposals

The above results proved that it is possible to assess different organizational structures according to different metrics. Beyond the analysis herein documented, the simulation of the real airline organizational structure makes it possible to evaluate other scenarios or introduce new metrics. As an abstract model based on reality, there is always room for simulation evolution.

References

Andersson, T., 2006. Solving the flight perturbation problem with meta heuristics. JoH 12 (37), 53.

Ball, M., Barnhart, C., Nemhauser, G., Odoni, A., et al. 2007. In: Barnhart, C., Laporte, G. (Eds.), Air Transportation: Irregular Operations and Control. Elsevier, Amsterdam, Handbook in OR & MS, 22.

Calvin, H.P., Pava.F P., 1984. Managing New Office Technology: An Organizational Strategy. Free Press, New York.

Castro, A., 2008. Centros de controlo operacional: Organizacao e ferramentas. Monograph for Post-graduation in Air Transport Operations. ISEC – Instituto Superior de Educação e Cîencias.

Castro, António J. M., Oliveira, Eugénio, 2010. Disruption management in airline operations control – an intelligent agent-based approach. In: Zeeshan ul-hassan Usmani PhD (Ed.), Web Intelligence and Intelligent Agents. INTECH.

Clancey, W.J., 2002. Simulating activities: relating motives, deliberation, and attentive coordination. Cogn. Syst. Rev. 3 (3), 471–499.

Clarke, M., 1998. Irregular airline operations: a review of the state-of-the-practice in airline operations control centre. J. Air Transp. Manag. 4, 67–76.

Clausen, J., Larsen, A., Larsen, J., Rezanova, N., 2010. Disruption management in the airline industry – concepts models and methods. Computers & OR 37, 809–821.

Greenbaum, J., Kyng, M., 1991. Design at Work: Cooperative Design of Computer Systems. Mahwah, New Jersey, Lawrence Erlbaum Associates.

Grosche, T., 2009. Computational Intelligence in Integrated Airline Scheduling. Heidelberg, Springer-Verlag, Berlin.

Guo, Y., 2005. A decision support framework for the airline crew schedule disruption management with strategy mapping. In: Operations Research Proceedings. Springer-Verlag, Heidelberg, Berlin.

Kohl, N., Karisch, S., 2004. Airline crew rostering: problem types, modeling, and optimization. Ann. Oper. Res. 127, 223–257.

Kohl, N., Larsen, A., Larsen, J., Ross, A., Tiourine, S., 2007. Airline disruption management: perspectives, experiences and outlook. J. Air Transp. Manag. 13, 149–162.

Liu, T.K., Jeng, C.R., Liu, Y.T., Tzeng, J.Y., 2006. Applications of multi-objective evolutionary algorithm to airline disruption management. IEEE International Conference on Systems, Man and Cybernetics, New York.

Mayer, R., Benjamin, P., Caraway, B., Painter, M., 1998. Framework and a suite of methods for business process reengineering. In: Grover, V., Kettinger, W.J. (Eds.), Business Process Change:. Idea Group Publishing, Reengineering Concepts, Methods and Technologies.

Nissen, R., Haase, K., 2006. Duty-period-based network model for crew rescheduling in European airlines. J. Sched. 9 (255), 78.

Sierhuis, M., 2001. Modeling and simulating work practice. Brahms: a multi-agent modeling and simulation language for work system analysis and design. PhD Thesis, Dept. of Social Science Informatics, University of Amsterdam, Amsterdam.

Sierhuis, M., Clancey, W., 2002. Modeling and simulating work practice: a method for work systems design. IEEE Intelligent Systems 17 (5), 32–41.

Wei, G., Yu, G., Song., M., 1997. Optimization model and algorithm for crew management during airline irregular operations. J. Combin. Optim. 1 (305), 21.

Yu, G., Qi., X., 2004. Disruption Management: Framework Models and Applications. World Scientific Publishing Company.

CHAPTER 7

A Multi-Agent System to Study the Internal Displacement of Passengers and Their Distribution on a Large-Capacity Bus

Antonio Neme*, John Graham*, Sergio Hernandez**, Omar Neme†

*Complex Systems and Non-linear Dynamics Group, Universidad Autónoma de la Ciudad de México, San Lorenzo, Del Valle, México, D.F. México; **Postgraduation Program in Complex Systems and Non-linear dynamics, Universidad Autónoma de la Ciudad de México, San Lorenzo, Del Valle, México, D.F. México; †School of Economics, Instituto Politécnico Nacional, México, D.F. México

7.1 Introduction

Since the seminal works of Helbing and Molnár (1995) and Batty (2003), the behavior of pedestrians has been the subject of study by the traffic and urban modeling community, and in particular, from the agent-based models perspective. The behavior and dynamics of pedestrians are not trivial to model and in many cases they are more complex than the dynamics shown by vehicular traffic.

Although vehicular traffic is constrained by lanes and directions of flow, and laws pertain that drivers more or less accept, pedestrian traffic is comparatively lawless and its displacement more or less unconstrained (Jian et al., 2005; Rangel-Huerta and Muñoz, 2010).

In general, there is a traffic law that is more or less accepted by drivers, who are constrained by the considerations of the traffic regulations (Bazzan and Klugl, 2009). However, such law does not exist in the case of pedestrians, who have more liberty in their displacements. Several aspects of pedestrian dynamics have been studied, including intermingling, routing (Papadimitriou et al., 2009), free flow, and behavior under panic (Helbing and Molnár, 1995).

A special case of pedestrian traffic is that observed in confined spaces, such as in galleries or in malls (UAS, 1996). In these cases, pedestrians follow a path determined by the collection distributions (or stores) and the density of other pedestrians (Batty, 2005, Batty, 2003). In the confined space case, there is one case in particular that has not been extensively studied, and has a big impact on daily life: the pedestrian (passengers) dynamics inside high-capacity public buses.

Advances in Artificial Transportation Systems and Simulation.
Copyright © 2015 Zhejiang University Press Co., Ltd. Published by Elsevier Inc. All rights reserved.

Public buses in some cities are normally large enough to contain up to 200 passengers, some standing and some seated. Normally, the journey is limited to the city limits, and the time necessary to cover the whole route may vary from a few minutes to a couple of hours. Normally, passengers do not travel for the entire journey, and there is an exchange of passengers at intermediate stations (Salazar and Lezama, 2008). These exchanges lead to nontrivial dynamics, as some passengers tend to move to more comfortable areas, whereas others tend to stay in areas close to the exit. In some high-capacity buses, the behavior of users may be affected by physical constraints such as seat distribution, entrances, exits, and handrails. Also, the overall dynamics observed in the bus are affected by the strategies followed by users in order to find a seat or a more comfortable position. Our interest is in the dynamics followed by passengers in a high-capacity bus, in particular, that observed in the *Metrobus* public transport system in Mexico City (referred to as MB). Here, we will refer constantly to this particular public transport system, as much of the data was obtained from studying it.

In the MB buses, users tend to stay close to the entrance/exit facility, apparently without considering journey duration or the availability of space in other areas. This situation leads to uneven passenger density distribution in the bus, as some areas are not visited by passengers, whereas other areas, mainly those in the exit neighborhood, are very dense. We refer to this distribution as heterogeneous and it is the cause of discomfort, as passengers see the available areas with lower densities reduced by blocking. Also, as the doors of the MB tend to be blocked, users have to make a major effort to pass through the denser areas in order to leave the MB.

There are a number of explanations for this behavior. Most of them are only observational and have not been quantified. One common explanation is that passengers know the bus will achieve the maximum capacity and moving around in that situation to get to the exit is not an easy task (Antonini et al., 2006; Was, 2010).

Currently, the system of entry to the MB is regulated from the platform, which is constructed with different portals that match the location of the bus doors and are assigned to separate passengers over 65 years old and women from men at the point of entry to the bus. Everyone, both men and women may exit through any door. We were interested to test if this scheme is optimum in the sense of homogeneous distribution and comfort for passengers, as defined in section 7.3. Indeed, we give evidence from simulations that if the entrance door were differentiated from exit doors then the distribution of users in the bus would tend to be more even and some measures of comfort would be higher.

The problem of particular interest to us is that of heterogeneous distribution of users inside the bus, which impacts on the quality of the trip from the perspective of users. There are three main perspectives in the study of pedestrian dynamics. The first one is continuum dynamics, in which pedestrians are seen as a fluid and physical rules defined as social forces are considered (Scovaner and Tappen, 2009; Treuille et al., 2006). Some variations consider the so-called "thinking fluids," where fluids standard dynamics arc adapted (Hughes, 2003). The second perspective is that of cellular automata, where time and states are continuous and a transition function dictates the behavior of pedestrians (Burstedde et al., 2001).

The third one is the agent-based model. We present an agent-based model to study the users' dynamics once they enter the bus, trying to capture the behavior shown by users (obtained from surveillance videos). Also, we analyze other policies to enter/exit the bus that may lead to an increase in the journey quality.

Agent-based models are a relevant tool to study pedestrian traffic as they keep a complete record of behavior, visited locations, and any other variable, at the time that agents are allowed to adapt their behavior (Antonini et al., 2006). Thus, we constructed an agent-based model to study the dynamics inside the MB and to try to identify policies that may lead to more comfortable distribution of passengers inside the bus. The model we propose falls in the category of microscopic models of passengers' interactions.

In an interesting work by Was, the concept of proxemics, that formalizes the spaces in which people interact with both peers and the environment, has been applied to study crowd dynamics (Was, 2010). The studied dynamics include the cases of free-flow and panic situations, in which well-defined areas surrounding people are made inaccessible to other pedestrians depending on the situation. These areas are intimate, personal, and public spaces and other pedestrians are not allowed to enter if the stress level is low. The space use and pedestrian dynamics shown by Was are related to the case of crowded areas in general, and may be related to crowds in confined spaces, such as in public buses. The concept of discomfort we apply in this contribution is a generalization of the one presented in (Was, 2010).

As an extension of cellular automata, Bandini et al. (2007) propose the study of pedestrians in a situation similar to the one proposed here, the dynamics of users inside a bus. The model they propose entails confining agent movement to discrete places and in an order that, while perfectly compatible with the agent-based methodology they have successfully formalized, would not, we consider, be the best option in the real-life circumstances of MB travel as collisions may be undetected and different sizes of agents could not be represented. Helleboogh et al. (2007) present a perspective of modeling dynamic environments in the context of multi-agent simulation. There, authors emphasize the use of particular methodologies to formalize the dynamic environments, as the one observed in the MB.

Our agent-based modeling has attended to the size of the bus, its passenger capacity, seats and entrances. As the dynamics in the interior are affected by the incoming flow of passengers, we followed an entrance/exit distribution of passengers based on *in situ* observations over the total number of stations of one of the MB routes. Then, we tested several scenarios in which the policies for entering and leaving the bus are changed and also we modified the distribution of inner spaces, by changing the location of seats.

In Section 7.2 we introduce the agent-based model of passengers, which include the general features of the MB system. In Section 7.3 we show the simulations, experiments and scenarios considered for the study of the internal dynamics of the Metrobus and the main results. Finally, in Section 7.4 some conclusions are stated.

7.2 The Model

We aim to understand the dynamics followed by agents when the rules of action lead to situations in which passenger density is heterogeneous in the bus and general discomfort is high. This understanding may allow us to explore alternative policies that may exert some influence on users to change their dynamics and avoid the observed situation in which there are some crowded areas at the same time as there are areas with very low densities.

In order to study a reduced version of passenger dynamics in a high-capacity bus, we developed a situated multi-agent model of the bus and the passengers. The model is a hybrid architecture similar to that introduced in (Shao and Terzopoulos, 2007; Vizzari and Olivieri, 2009), in which the world and its components are described in terms of states, rules, and equations. Our model consists of three components: agents, physical environment, and external environment.

The first component of the model refers to passengers and their attributes, internal states, and rules that guides their behavior. The second component is the representation of the physical environment, which includes seat distribution, entrances/exits locations and sizes, as well as available areas and internal corridors. The third component of the system is related to the observed number of passengers entering/leaving at each station of the MB network, as well as the traveling time between stations. We call this third component the external environment.

In the following subsections we detail the agent-based passenger component, the internal physical environment and the external environment. It is important to note that in our model we do not consider the presence of passengers traveling together as a group. Those groups have been studied and found to be relevant for crowd dynamics (Manenti et al., 2011). However, the presence of groups is very uncommon in the system we are studying (EOD, 2007).

7.2.1 Agents and Passengers

The passengers inside the MB are represented by agents. Each agent is represented as 3-tuple $<s, p, r>$, where r is the agent type, p is the region in space the agent is situated in, and s is the state for agents. Besides states, agents have a set of attributes that reflect both the variables a passenger may take into account, and his/her particular behavior (see Figure 7.1). The space in which agents are allowed to move is continuous (no discretization other than the precision of floating points).

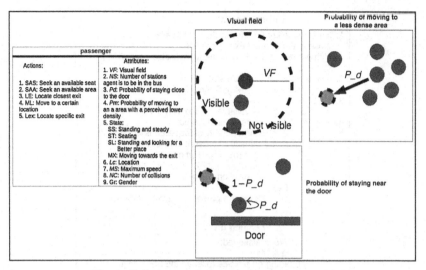

Figure 7.1: Agents representing passengers.
(a) Agents may follow five actions, present one of five states, and are described by nine attributes. (b) Description of visual field VF and probability of moving to other regions. (c) The state transition diagram

The first attribute is the visual field of the agent, defined as the circumference of radius *VF* that the agent may explore. The second attribute is the number of stations the agent will travel (see Section 7.2.3). *P_d* is interpreted as the tendency of the agent to stay close to the door even when there are available seats or enough room in other areas of the bus interior; and where *P_m* is the tendency to move to areas in which passengers' density is lower. Gender is also a relevant attribute for passengers, as some policies for entering and leaving the MB are based on it.

There are five actions that agents can take. Once agents enter into the MB they can seek an available seat (SAS) within their visual field. This action is not followed by all agents, as it depends on the tendency of agents to stay close to the door (*P_d*). The second action agents can follow is to seek an available area (SAA). Agents that try to find a seat and are not able to do so will try to find an available area. Once agents find a seat or an available area they move to that location, which is the third action (ML). Once agents are about to arrive at their desti-nation, they either locate the closest exit (LE) or locate the exit from where they should leave the MB (Lex) and move to that exit (ML).

When agents perceive their environment (which is limited by their own visual fields), they may take the decision of moving to another region within. That is, the environment and the internal state will lead agents to a new state and also may lead them to effect an action: s, p, r *Environment* $(VF) \rightarrow s', p', r'$ *Action*. Actions are of two kinds: (1) displacement (*ML*) and (2) visual inspection (*SEK, SAA, LE, Lex*). Both kinds of actions implicitly include a

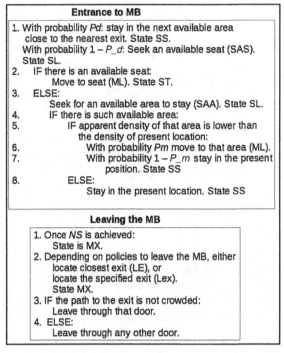

Entrance to MB

1. With probability Pd: stay in the next available area
   close to the nearest exit. State SS.
   With probability $1 - P\_d$: Seek an available seat (SAS).
   State SL.
2.   IF there is an available seat:
         Move to seat (ML). State ST.
3.   ELSE:
         Seek for an available area to stay (SAA). State SL.
4.   IF there is such available area:
5.   IF apparent density of that area is lower than
             the density of present location:
6.   With probability Pm move to that area (ML).
7.   With probability $1 - P\_m$ stay in the present
                 position. State SS
8.   ELSE:
             Stay in the present location. State SS

Leaving the MB

1. Once NS is achieved:
   State is MX.
2. Depending on policies to leave the MB, either
   locate closest exit (LE), or
   locate the specified exit (Lex).
   State MX.
3. IF the path to the exit is not crowded:
   Leave through that door.
4. ELSE:
   Leave through any other door.

Figure 7.2: Main algorithm for agents representing passengers

cognitive process such as planning and collision avoidance for *ML* and distance estimation for the visual ones. When agents intend to change position, either because a seat is available, a less crowded area is visible, or the number of stations *NS* has been reached, they will try to follow a straight line between the actual position p and the new calculated position p' (represented as an arrow in Figure 7.1), applying a collision-avoidance mechanism to avoid other agents and seats.

As agents take actions, so their states are shifted. Also, in order to take certain actions, their own present states are taken into account. The main algorithm that reflects the actions for agents is shown in Figure 7.2.

When moving to the selected position, each agent tries to avoid collisions with other agents and with seats. Agents maintain a record of the number of collisions and quasi collisions experienced during their trip (*NC*, see Section 7.3). Agents standing in an area within the neighborhood of another agent, or moving at a low speed, do not allow the latter to move across them. Thus, the agent has to explore other paths to get to the desired location (exit, seat, available area). Once agents have reached their stop, that is, *NS* has been achieved, the agent moves either to the next available exit or to the specified exit, and if it is not crowded, leaves the bus from here. If it is crowded, then the agent will try to reach the next available exit.

Agents may be in one of four states: (1) standing and steady (SS), (2) sitting (ST), (3) standing and looking (SL) for a better place or an available seat, and (4) moving toward the exit (MX) (see Figure 7.1). The agents follow the established rules and according to the environment can change states. It may happen that some agents do not visit all states, as, for example, some agents will never be able to find a seat or others will remain standing and steady near the door. States SL and ST are the less frequent states, affected by the attributes P_m and P_d but also by the density and number of passengers in the MB.

Following the mentioned rules (see Figure 7.2) and from the individual attributes and actions of agents, they interact inside the bus. The number of passengers varies as a function of the external environment that we define in Section 7.2.3, and as passengers enter and leave the bus in each station, the number may vary following any distribution.

When an agent is trying to move to reach a seat or an area with lower density, or the corresponding exit, he/she may find the way blocked or may collide with other agents. In such situations, each agent will randomly choose to move either to their left or right, and try to continue their displacement. This behavior is also observed in real-life situations.

Once a critical number of users are inside the bus, available space is constrained and crowded dynamics emerge. Those dynamics are in general a cause of discomfort and a source of heterogeneous density of users within the available space, that is, blocking of paths to exits/entrance doors.

All attributes may be selected individually for each agent. In Section 7.3, we describe how we defined these attributes. Agents are represented only once they enter the MB through a prespecified door. The entrance door for every agent is dictated by the established policies. In Section 7.3.1 we describe a number of experiments in which entrance policies are modified. The exit door may also be determined by those policies, but in general, if the predefined exit door is not reachable because of a crowding situation, agents may leave through any door.

7.2.2 Representation of Internal Physical Environment

The physical environment is formed by the interior of the MB. It consists of doors, seats, a corridor, and standing areas. In general there is only one standing area, defined as a rectangle, but in order to give more flexibility to our model so as to consider some kinds of bus architectures, several areas are possible (see Figure 7.3b). In the standing area, several doors and seats can be placed. In the model we constructed, several aspects of the physical environment of the MB may be specified. Figure 7.3 shows the structures that may be specified. Each component of the physical environment is represented as an agent that defines their size and location, as well as state.

The areas in which pedestrians move tend to be modeled like conducting fluid in pipes or as a discrete structure (Schadschneider et al., 2009). The former case applies more naturally when

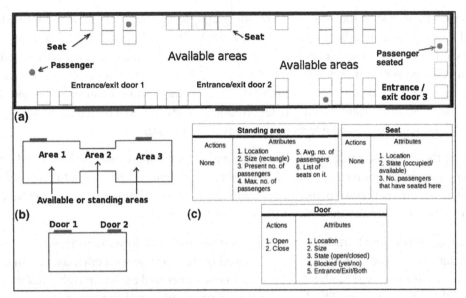

Figure 7.3: The MB physical environment.
(a) The structure present in the MB in Mexico City. (b) Examples of possible distributions.
(c) Attributes of physical environment agents

force-based models are implemented whereas the latter is better suited for situated agents (Ferreira, 2009). Cellular automata (CA) are a common option in modeling areas and pedestrians (Bandini et al., 2007). However, the tessellation obtained in a discrete space may not be adequate in the case we are studying for a number of following reasons:

1. Passengers may collide with others. In CA, the way to register collisions is that of counting occupied neighbor cells, which is unable to differentiate between real collisions and standing neighbors.
2. Passengers move with distinct speeds and have different sizes, which complicates the use of a discrete space.
3. Density measures are more accurate in a continuous (or quasi-continuous) space.

In this model, the space is continuous, and agents occupy a given region of that space. As agents may have different sizes, it is more natural to define the space on a continuous basis. For example, seats may be of different sizes, passengers may also have different sizes, and the size of a seat may be different from the size of an agent.

Seats have an orientation that simply reflects how the passenger may sit on it. Passengers should only have a seat with access from available sides, and not through the backrest. There are four possible orientations to seats: (1) forward, (2) backward, (3) pointing to the right (backrest toward the left side of the MB), and (4) heading to the left (backrest toward the right side of the MB). In the graphical representations, the orientation is shown by a color code. If no color is shown, the default (forward) orientation is meant. Seats may be placed anywhere in the available area, and passengers are not allowed to pass through them. This is an important fact, as seats may serve as blocking barriers, as described in Section 7.3.2. The size of the seat may be defined by the user, but in this work we fixed it as the equivalent of a square of $0.5 \times 0.5 \, \text{m}^2$.

Doors may only be located at the boundaries of the available area. Doors may be for entrance, exit, or both. Size of the doors may also be defined by the user, although in this work we fixed the size of the three doors to reflect the distribution in real MB. At least one door should be present in the model, but there may be any number of seats, or none, in an available area.

7.2.3 External Environment

The dynamics of passengers are affected by the number of passengers in their surroundings. If the number of passengers traveling in the MB is very low, then no relevant distributions would be present. The number of passengers that enter or leave the MB at the stations of the route is then a relevant feature. In our model, it is possible to specify the number of stations, the number of passengers that will enter the bus at that station and also the number of passengers that will leave the bus at it.

It is common to assume a Gaussian distribution for the number of passengers entering or leaving in any station. However, in our system it is possible to specify the number of passengers entering/leaving the MB in two ways: (1) By a probability distribution function. (2) By directly specifying for each station the number of passengers that will enter/leave the MB.

The observed passenger dynamics during the trip are also affected by the distance between stations. Stations located very close to each other have different effects to stations located far away from each other because the available time for passengers to try to look for better places is longer. In our model, it is also possible to specify the duration between any pair of consecutive stations.

Figure 7.4 shows the information that defines the external environment for MB. It shows a segment of a real network and the relevant data about the number of passengers entering and leaving the MB.

The model we present here consists of three components: (1) passengers, (2) the physical environment, and (3) the external environment. The dynamics observed in the bus interior are the result of the influences of attributes of these three components, and of the attributes of the physical environment defining passenger behavior in which agents can be situated. Passenger density is also relevant and is determined by both, the external environment and the rules followed by agents. In the next section we describe the set of experiments and simulations we conducted in order to try to understand the dynamics observed in the interior of the MB.

The software was implemented in Python, using library Pygame (software is available from the authors). The user interface to specify the three components is straightforward. A whole new virtual network can be specified in a couple of minutes.

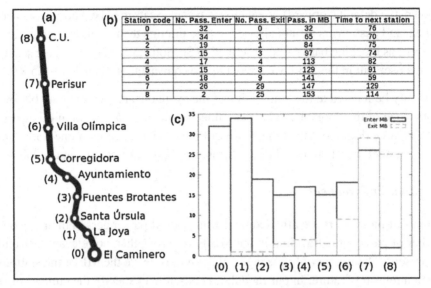

Figure 7.4: External environment specifications.

A route (a) is specified in terms of the number of passengers that enter and leave the MB in each station and in terms of the traveling time between the station and the next one (b). In (c) is shown the number of passengers that enter and leave the MB in each station. (Information collected by the authors)

7.3 Simulations and Results

In the proposed model, each passenger is represented by an agent, with his/her own features and dynamics, and displacements are in a parallel mode. Also, the physical environment is established according to a set of attributes (see Section 7.2.2) and the external environment is also specified. All these attributes are the control parameters and may be specified individually or following a given distribution. The order parameters (or observable variables) present a summary of the behavior of a system.

The agent-based model to study the dynamics and behavior was developed in Python. Graphical interfaces are part of the model. Figure 7.5 shows a snapshot of the main screen. Agents representing females moving toward a seat or to a better area are represented as dotted light-gray circles. Females seated are dotted light-gray circles with a horizontal line and females standing steady have a vertical line in the middle. Males moving to seats or toward other areas are white circles. Males standing steady are white circles with an inner dot and males seated present an inner cross. When agents are moving toward the exit, they change their gray scale: to a darker tone.

As mentioned, in our case study we are interested in the heterogeneous distribution of passengers inside the bus. The order parameters we identified as relevant to summarize those order parameters are, besides the passenger density in the available areas, two measures of comfort.

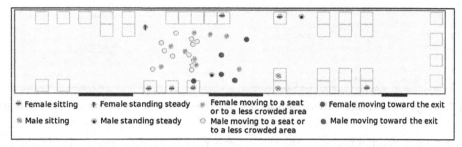

Figure 7.5: Snapshot of the agent-based model.
Agents are differentiated accordingly to their state. Agents are colored accordingly to gender and if they are moving to leave the MB or if they are moving to find a better place. Agents seated present a horizontal bar (females) or a cross (males). Those agents standing steady are also differentiated with a vertical bar (females) or with a dot (males)

The first one is the number of *quasi*-collisions (*Q*), and the second one is the perceived discomfort. The first comfort measure counts the number of times the trajectories of two passengers are intersected. The second measure is known as discomfort, and is defined in equation (7.1):

$$Ds = \frac{1}{N} \sum_i^{agents} H(i,k) \qquad (7.1)$$

where $H(i,k)$ is the number of agents in a circumference with the center at the position of agent i and radius k. It is a measure related to the density and is the average over all N passengers that traveled in the MB. High-density areas are the main source of discomfort in journeys in public transport (EOD, 2007).

Even when a few passengers tend to stay in regions near the exit (low P_d), they may be enough to block other areas and thus the passengers that enter the bus will not reach those areas with low density. As a consequence, they will stay in areas close to the exit, not from choice, but out of necessity, as passage to other more comfortable areas is blocked by other passengers (see Figure 7.6).

Trying to modify directly the behavior of passengers is not an easy task. A good solution to try to avoid heterogeneous distributions and blocking in the MB could be to forbid passengers to stay near the doors and try to convince them to move in an ordered fashion to areas with low densities. However, this modification of the behavior implies a change in cultural backgrounds, which is not an easy task.

The only alternative to try to make the journey for passengers more comfortable is to modify the physical environment. By this modification, we refer to modifying the actual schemes for entering and exiting the MB. Also, modifications in the seat distribution were also tested. Two sets of experiments were developed in order to study the changes in the order or observable parameters (comfort and passenger density). In the first experiment, the entrance/exit

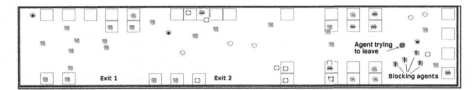

Figure 7.6: Agents that block the free transit toward available seats or areas with lower density

policies were modified, whereas in the second experiment, modifications were made to the seat distribution over the MB.

We carried out several experiments to study the dynamics of passengers. The attributes of agents were modified in order to reconstruct the observed behavior. That is, a small number of passengers decide to stay close to the exit and, as a side effect, they obstruct the free entry to the bus, thereby generating a high density of passengers at that point of entry by impeding movement to other regions with probably less density. We modified the parameters shown in Figure 7.1 to reproduce that situation. We calibrated the system on an empirical basis, from data achieved from observations *in situ* and from video surveillance (Videos). In particular, we settled $P_d = 0.02$ as at this value, and the observed blocking in areas surrounding the doors became present. The rest of the control parameters we defined to follow what is reported in (EOD, 2007), and in (Batty, 2003). In addition, modifications to the environment were studied, such as differentiated enter/exit schemes, in order to quantify their impact on the passenger distribution and in general user comfort.

7.3.1 Modification of the Entrance/Exit Policies

Once the observed behavior was reproduced in the model through a tuning of the control parameters, we modified agent's rules in order to identify a situation in which collisions are reduced, and at the time that user distribution in the bus is not as odd as in the present case, in which areas close to the exit doors tend to be crowded even when passengers will travel for a long distance.

Four scenarios were considered. Case A is that of total freedom: passengers may enter and leave the MB through any door, and with no gender differentiation. In case B males and females now have different options. Males may enter only through doors 2 or 3, and may exit through doors 2 or 3. Females may enter the bus mainly through door 1 (probability of entering the bus through this door is fixed to 0.9) and may leave the MB through any door. Case C makes no distinction between males and females: all passengers must enter through door 2 (the central one) and may leave through doors 1 or 2. Case D is again the same for women and men, and both can enter the bus through any door but the constraint is that they should leave the MB through the same door they entered by.

The four scenarios were tested over 100 runs, and each run represented a whole trip that covered all 30 stations, and with the external environment (see Section 7.2.3) consistent with observations. The results shown here are the average over the 100 runs. Case B is the observed policy in the real MB in Mexico City. It was implemented to minimize interactions between males and females, as a number of misconducts had been reported when the bus was crowded. For the four scenarios, the same conditions were maintained in terms of passenger features, the external environment, and the internal environment. The only variations were those regarding entrance and exit doors. When there are no constraints about the entrance door, agents decide randomly the door by which to enter the bus. When there are no constraints about the exit door, agents try to leave through the closest one at the moment they approach their stop. Once agents enter the MB through the prespecified door, both males and females may move to any part of the bus following their attributes and rules. The actual policies tend to keep apart males and females inside the MB, but in general those regulations are not followed.

The number of agents that enter and leave the bus was obtained from a near-peak hour schedule of a working day for a 16 km journey that includes 30 bus stations (EOD, 2007).

Figure 7.7 shows the average density of agents that have been standing and steady in each area in the bus. As was expected, the areas close to the exit/entrances present the highest

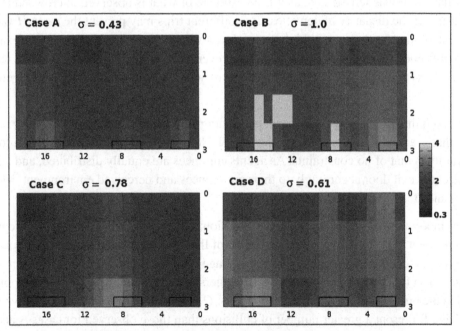

Figure 7.7: Average passenger density in standing and steady state (SS) over each area of the bus for the four scenarios (cases A–D).
Case B presents the highest differences in distribution, whereas the no-constraint case presents the lowest differences. For case B, the areas with the highest density are those near the doors with the number of steady agents. Doors are shown as rectangles in the bottom side of the MB

number of passengers. These results were achieved with $P_d = 0.02$ and the measure was taken after MB left each virtual station. Areas near seats tend to present high densities, as agents will move simultaneously toward seats once they are available. As only one agent can be seated, all agents that moved to have a seat will remain in the vicinity. This behavior is also observed in the real MB. In order to compare densities in the four scenarios, we defined a measure that takes into account the heterogeneity in the densities in the available area (equation 7.2):

$$\sigma(H) = (M_H - m_H) / M' \tag{7.2}$$

where H is the average density of standing and steady passengers for each region of the MB. That is, H is a matrix whose rows and columns are the position of each available area and values at position *(i, j)* of H are the average number of standing and steady agents in that square. The available areas are squares that account for approximately 1 m². M_H is the maximum average number of standing and steady passengers and m_H is the minimum average number of steady passengers. Both quantities refer to simultaneous events. That is, we are measuring the differences in density at every time and recording only the highest one. M' is a normalizing factor and is the maximum number of agents that may be standing in a given area. In our simulations it was settled as 5, as it is an estimate of what is observed in crowded trips. H is a measure of the disparity of densities. Two different trips may present the same H value, even if one of the trips has a very high discomfort level and the other a low one. For example, a group of friends may be traveling together and even in an empty MB may decide to remain standing in a very narrow area, which may contrast with a trip in which there are few empty areas.

It is observed that the density of passengers standing steady near the doors is higher in case B, the one that is actually implemented in the MB of Mexico City. The scenario with the lowest dissimilarity is that of no constraints. As agents entrances are equally distributed, and agents may select the exit door accordingly to their preferences and perceived environment, less standing and steady agents are found.

Another important measure is that of quasi-collision, Q. It occurs when two agents are moving and they get very close to each other, so one of them has to recalculate the path to arrive at the desired location. Figure 7.8 shows the average number of near or quasi-collisions for each agent as a function of the number of passengers that are standing (either moving or steady) in the bus. It is observed that for all cases, Q grows more or less linearly. However, cases C and B present a greater number of collisions than those observed for cases A and D.

The results obtained by the agent-based model resemble those present in the bus. In both cases, there is an odd density of passengers, that is, some areas present a high number of users whereas others are almost empty. This feature is not desirable in public transportation, as the bus does not reach its optimum capacity at the time users tend to feel uncomfortable.

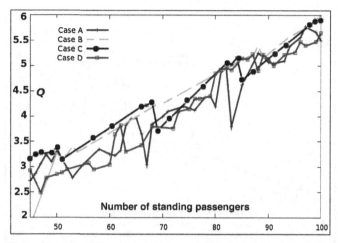

Figure 7.8: Number of quasi-collisions (Q) as a function of standing passengers in the bus.
Case B presents the smoothest curve, although the number of collisions for this case tends to be
greater than for the remaining cases

Furthermore, the areas with higher density are those closest to entrance/exit doors, which blocks passengers trying to enter the bus.

The observed density distribution in case B (the actual scheme) is caused, at least in part, by the lack of users displacement once inside the bus. Once they find a free space, they move toward it and try to stay the whole journey in it. This free space may be located very close to the entrance door, which affects the free flow of users trying to enter at subsequent stations. By forcing the users to leave the bus through a door different from the one they entered (cases A and C), it is observed that users tend to move and thus free spaces are not available near entrances.

In our simulations, the number of agents and the distance traveled is adjusted to the available data (EOD, 2007). The number of agents that enter each station and the number of stations for each travel was approximated to the origin destination matrix from the most recent questionnaire (2009) (MOD, 2008; PTV, 2008). For the considered scenarios, the number of users and the matrix were the same.

As the perceived discomfort Ds is a measure of user density, we continued to use it to compare the four considered scenarios. In Figure 7.9, it is observed that the discomfort Ds is higher for case B than for any other case. Also, it is observed that C grows smoothly for case B, which contrasts with what is observed mainly in cases A and D, in which there are some bursts present.

Agents do not necessarily leave the bus through the same door that they selected to enter. This fact is a measure of mobility. Figure 7.10 shows the average probability that agents will leave the bus through a door different from the one they entered the bus by (P_E), as a function of the

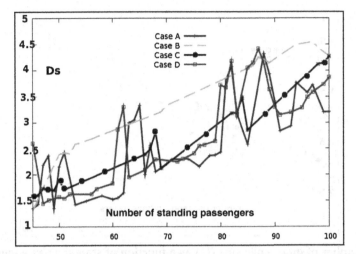

Figure 7.9: Discomfort (Ds) as function of the standing passengers.
It is the average number of agents surrounding each agent within a distance $k = 1.0$ m. In all cases, the number of agents in the bus is the same for each station. It is observed that in general case A presents the lower discomfort, whereas case B is the worst of the studied policies

number of standing passengers on the bus at that moment. It is observed that, as more agents are in the bus, the probability that a passenger will leave the MB through a different door increases. Case D is an exception, as agents are required to leave the bus through the same door as they entered by. That is, *PE* is almost 1.

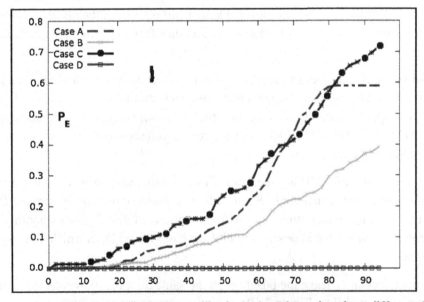

Figure 7.10: Probability that a given agent will exit the bus through a door different from the door by which he entered the bus, as a function of the number of standing passengers in the bus

For case A, we observe what seems to be a second-order phase transition. As the number of passengers increases, it gets harder for agents to move to another region, as the passengers may block the available paths. Passengers *en route* to the exit will find their way blocked and thus will need to change their paths, leaving the MB through a door different from the one by which they entered the MB. Figure 7.10 was obtained for 200 runs, varying P_d from *0.01* to *0.05*.

Case B presents a slow growth curve. When the MB has reached its maximum capacity, more than half of passengers will still leave through the same entrance door. This proportion may indicate a lack of mobility, also present for case C, which shows a curve with more or less sustained growth. For case C, only when the maximum bus capacity is about to be achieved, is it observed that this pattern changes.

7.3.2 Modification of Seats Distribution

The second option for improving the comfort measures is to modify the seats distribution. To do so we tried to find a seat distribution that in the same circumstances led to a better performance in terms of passenger comfort. In this experiment passenger behavior and external environment were fixed, as in the previous experiment, and the four possible scenarios detailed in Section 7.3.1 were evaluated as well.

A genetic algorithm was applied to find an optimal seat distribution. Genetic algorithms are a search heuristic based on evolution by means of natural selection (Mitchell, 1998). Here, individuals in a population represent solutions to a given problem. By means of a biased selection of good solutions over the rest, and by an exchange of parts of the solution, the algorithm is able to find adequate individuals that represent good solutions. Here, solutions consist of seat distributions. The objective function to minimize is given by equation (7.3):

$$OF = (Ds + Q / K + \sigma(H))3 \tag{7.3}$$

where Ds is the comfort averaged for all agents for the whole journey (see equation 7.1), Q is the average number of collisions for all agents, K is the maximum number of collisions an agent can experience (preset to 10), and $\sigma(H)$ is the difference between the maximum number of passengers standing at the same time in an area and the minimum number of passengers standing in another area at the same time (see equation 7.2). Figure 7.11 shows the general algorithm followed to obtain seat distributions with the highest possible comfort. For all cases, the external environment and passengers attributes were the same, as were also the attributes of the doors (location and size). Also, the number of seats was fixed, so the only free parameter was the position of seats and their orientation. Each epoch takes account of the whole simulation for the 30 stations schedule, each one with the same external variables (see Section 7.2.3).

Seats may be located at any position as long as they do not overlap with another seat, and as long as they are not located in front of a door and at a distance from it of less

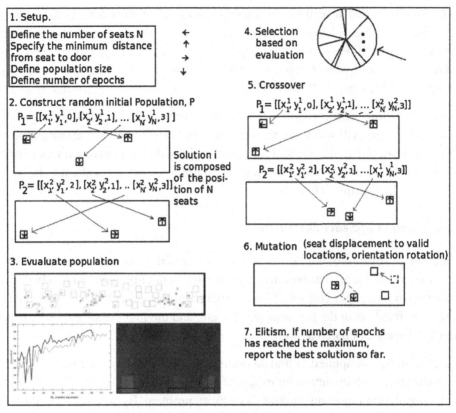

Figure 7.11: Genetic algorithms for the identification of the seat distribution that maximizes comfort.
Population size, 100; probability of crossover, 0.9; probability of mutation, 0.05; epochs 250. Solutions (individuals) represent seat distributions. The number of seats was fixed at 36. Elitism was included

than one meter. The number of seats was fixed at 36 and the genetic algorithm ran for 250 generations, with a population size of 100, probability of mutation 0.05, probability of crossover 0.9, and it considered elitism. The genotype (solution) has the form $Sl = \left[[x_0, y_0, O_1], [x_1, y_1, O_1], \ldots, [x_{35}, y_{35}, O_{35}] \right]$ where $([x_i, y_i, O_i])$ indicates the top left corner of the seat and the orientation of seat i. The phenotype is the MB constructed following the seat distribution specified in the genotype. The evaluation of phenotypes follows the process described in Section 7.3. As seat size is fixed, no more information is needed to specify the seat distribution. Also, the orientation of every seat is decided by the algorithm, but if the orientation is not compatible with the seat position, it is changed by the algorithm both in the genotype and phenotype.

If a solution presents overlapping seats, then one of the overlapping seats is displaced randomly to a nearby position (and fixed there). The reason for doing this guided displacement

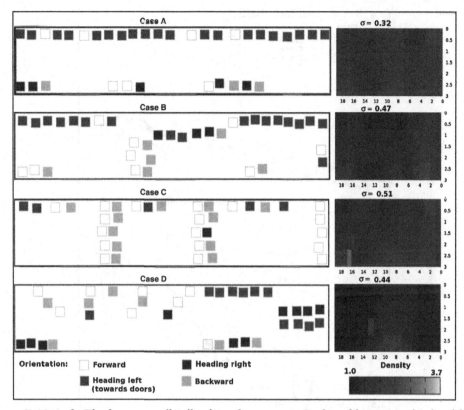

Figure 7.12: Left: The best seat distributions for cases A–D, found by a genetic algorithm. Right: the average passenger density

and not penalizing solutions with overlapping seats is that we are not interested in an algorithm able to learn how to displace nonoverlapping seats, but to find a seat distribution that maximizes comfort.

We present in Figure 7.12 the best solutions for each one of the four scenarios described in the previous section. The orientation code for seats is as follows: white seats are forward sets, gray means orientation to the left of MB, dark-gray represents orientation to the right, and light-gray represents seats facing backward.

In the four cases, clear patterns in the seat distributions were found by the genetic algorithm. For case A (no-constraints), the majority of seats were placed on the right side of the bus, opposite to the door entrance. For case B, it is interesting that the algorithm found a configuration that seems to compartmentalize the bus into two regions, one for males and another for females. The male section has two doors whereas the female section has only one. In case C, three independent sections were achieved, with one door assigned to each section. From the distribution for case D, the double row of seats near the back of the MB is outstanding. Also,

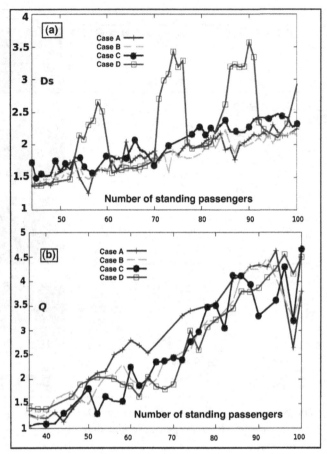

Figure 7.13: Ds and Q as a function of the number standing passengers.
Ds and Q for each one of the four policies and the associated seat configuration found by the genetic algorithm is displayed

the diagonal configuration near the middle part of the bus is interesting. The density distribution in all the four cases is less heterogeneous than in the configuration observed in the real MB (see Figure 7.7). $\sigma(H)$ is lower, as expected, in all the four cases.

In Figure 7.13 are shown Ds and Q for the best solution found by the genetic algorithm for each one of the four cases. Ds grow more slowly in the four new configurations than in the original one, and case D presents some interesting oscillations, that may need further analysis to be fully understood. For all four cases, Ds are lower than the counterpart (see Figure 7.9).

In Figure 7.13b are shown the number of collisions Q as a function of the standing passengers. As expected, Q is lower than the original counterpart, and Q grows more slowly as well. For all the four scenarios, the best-found seat configuration presents similar Q values. Only for case A is it slightly greater than for the remaining cases.

7.4 Conclusions

We have been able to reproduce in the agent-based model an equivalent situation, in which some passengers block the free transit in the available areas in a bus. It is not a planned behavior, but a consequence of the tendency of some passengers to remain close to the exit. Even if a small number of passengers follow this behavior, several areas may become inaccessible for passengers trying to get to them.

With the presented model, in which agents attributes and actions reflect to some degree the observed ones in real public transport users, we reconstructed some aspects that are also present in the studied phenomena. Among those aspects is the odd passenger density, but also the variation relating to the entrances and exits as a function of the number of passengers. That is, the higher the number of passengers in the bus, the more difficult it is to move around, and agents tend to stay close to the door they used to enter. Also, the number of quasi-collisions is an increasing function of the number of passengers in the bus. This behavior is present in high capacity buses.

In our model, we have modified the policies regarding the entrance/exit to the bus and we have shown that if some changes are implemented, the average quality of users may be increased, as the discomfort and the number of quasi-collisions is reduced when passengers are allowed to enter and exit the bus through any door. On the basis of reports of sexual misconduct, transport authorities have issued a number of laws that constrain males to enter through the same doors as females.

From the results here presented, we may conclude that the current policies for boarding and leaving the MB are not optimal in sense of comfort. A good policy will protect passengers at the time that offers users the most comfortable journey possible. Some alternatives may be implemented and agent-based models of those situations would be very relevant. In our model, the more comfortable situation corresponds to that in which entrance and exit doors are not differentiated, that is, users may enter or leave the bus through any door. Our model, however, does not contemplate the overall situation, in which users leaving the bus interact with users waiting on platforms. However, this situation may not be very relevant, as users on platforms may be retained a few seconds while users exit the bus.

If it is possible to modify the configuration or distribution, then it is possible to improve the comfort for passengers. For every particular policy for boarding and exiting the bus, there may be a seat distribution that minimized the discomfort as well as the number of collisions. Applying a genetic algorithm, we found a seat distribution that, for a given policy, improves the measures of comfort. It seems that a general policy may not be adequate for all seat distributions. It may be of interest for the public transport authorities to count with tools that are able to find seat distributions that improve, subject to a given policy, the average journey comfort.

Of course a number of extensions of this model are possible. From where will the improvements come and from what considerations would they be extended? In the context of this study, closer investigation of the real-life functioning of passenger behavior would disclose more of the variables of this dynamic. We have been analyzing a growing database of videos, and planning passenger survey interviews, as they could be incorporated in the investigative phase. The use of agent-based modeling would seek to incorporate those variables in the simulations that follow and those simulations would be directed toward solving the problems they analyze. This process implies strengthening creative relationships between disciplines Computer Science, Mathematics and Behavioral Sciences, as well as the close participation of traffic authorities.

Acknowledgment

This project has been funded by the Instituto de Ciencia y Tecnologia del Distrito Federal, under project PICCT08-55. A.N thanks SNI – CONACYT.

References

Anon Videos are available in the links: http://www.ci-sa.com.mx/, and http://www.modelos-predictivos.org.mx/MMBUS/videos/

Antonini, G., Bierlaire, M., Weber, M., 2006. Discrete choice models of pedestrian walking behavior. Transport. Res. B 40, 667–687, doi:10.1016/j.trb.2005.09.006.

Bandini, S., Federici, M.L., Vizzari, G., 2007. Situated cellular agents approach to crowd modeling and simulation. Cybern. Syst. 38 (7), 729–753.

Batty, M., 2003. Agent-based pedestrian modeling. Working paper No. 61 of CASA – UCL.

Batty, M., 2005. Cities and Complexity. Boston, MA. MIT Press.

Bazzan, A., Klugl, F., 2009. Multi-agent systems for traffic and transportation engineering. Information Science Reference. Hershey, PA.

Burstedde, C., Kirchner, A., Klauck, K., Schadschneider, A., Zittarz, J., 2001. Cellular automaton approach to pedestrian dynamics – Applications. arxiv:cond-math/0112119v1.

Helbing, D., Molnár, P., 1995. Social force model for pedestrian dynamics. Phys. Rev. E 51, 4282–4286, doi:10.1103/PhysRevE.51.4282.

Helleboogh, A., Vizzari, G., Uhrmacher, A., Michel, F., 2007. Modeling dynamic environments in multi-agent simulation. Auton. Agent Mult-Ag. Syst. 14 (1.), 87–116.

Hughes, R., 2003. The Flow of Human Crowds. Annu. Rev. Fluid Mech. 35, 169–182, doi: 10.1146/annurev.fluid.35.101101.161136.

EOD, 2007. Origin-destination survey in Mexico City. http://www.setravi.df.gob.mx/work/sites/stv/docs/EOD2007.pdf.

Ferreira P, Esteves E, Rossetti R, Oliveira E., 2009. Applying situated agents to microscopic traffic modelling. 108-123. In: Bazzan, A, Klugl F., (Eds) Multi-Agent Systems for Traffic and Transportation Engineering.

Jian, L., Lizhong, Y., Daoling, Z., 2005. Simulation of bi-direction pedestrian movement in corridor. Physica A 354, 619–628, doi:10.1016/j.physa.2005.03.007.

Manenti L, Manzoni S, Vizzari G, Ohtsuka K, Shimura, K. 2011. An Agent-Based Proxemic Model for Pedestrian and Group Dynamics: Motivations and First Experiments. Twelfth International Workshop on Multi-Agent-Based Simulation Taipei, Taiwan.

Mitchell, M., 1998. Introduction to Genetic Algorithms. Bradford Books. Cambridge, MA.

MOD, 2008. Yearly inform of activities. Metrobus. http://www.metrobus.df.gob.mx/transparencia/documentos/art14/XIX/Inf_consejo_anual09_1trim10.pdf.

Papadimitriou, E., Yannis, G., Golias, J., 2009. A critical assessment of pedestrian behavior models. Transport. Res. F 12, 242–255, doi:10.1016/j.trf.2008.12.004.

PTV, 2008. Yearly program for transport. http://www.ville-en-mouvement.com/ameriquelatine/telechargements/PITVNUESTRO.pdf.

Rangel-Huerta, A., Muñoz, A., 2010. Kinetic theory of situated agents applied to pedestrian flow in a corridor. Physica A 389, 1077–1089.

Salazar, C., Lezama, J., 2008. Construir ciudad, un análisis multidimensional para los corredores confinados de transporte en la Ciudad de México. Colmex.

Schadschneider, A., Klúpfel, H., Kretz, T., Rogsch, C., Seyfried, A., 2009. Fundamentals of pedestrian and evacuation dynamics. 124-154. In: Bazzan, A., Klugl, F., (Eds), Multi-Agent Systems for Traffic and Transportation Engineering.

Scovaner, P., Tappen, M., 2009. Learning pedestrian dynamics from the real world.

Shao W, Terzopoulos D. 2007. Autonomous Pedestrians. Graphical Models, 69(5-6): 246-274.

Treuille, A., Cooper, S., Popovic, Z., 2006. Continuum Crowds.

UAS. 1996. Tate Gallery, Millbank: A study of the existing layout and new masterplan proposal. Unit for Architectural Studies, Bartlett School. University College, London.

Vizzari G, Olivieri F. 2009. Towards Hybrid Situated Agents Based Virtual Environments. Web Intelligence/IAT Workshops 587-590.

Was, J., 2010. Crowd dynamics modeling in the light of proxemic theories. ICAISC 2, 683–688.

CHAPTER 8

Crowd Simulation Applied to Emergency and Evacuation Scenarios

João Emílio Almeida*, Rosaldo J.F. Rossetti*, Fábio Aguiar**, Eugénio Oliveira*

*LIACC, DEI, Faculdade de Engenharia da Universidade do Porto, Porto, Portugal; **MIEIC, DEI, Faculdade de Engenharia da Universidade do Porto, Porto, Portugal

8.1 Introduction

Efficiently managing and organizing crowds in emergency situations, having its origins in a fire or other hazard, has become an important area of study and research in the last years (Cordeiro et al., 2011). It represents an important role in the design of a building or urban planning. Practitioners are eager to use pedestrian simulators allowing them to predict crowd movements in normal and emergency situations, allowing them to create and test different solutions for different scenarios, as well as to test hypothesis, in order to minimize possible accidents' damages in the future.

The development of pedestrian simulators has emerged from the aforementioned need of managing crowds safely as well as predicting their behavior. Other applications include traffic simulation (pedestrian and vehicles interaction) and the creation of artificial societies for games or movies. Whatever is the reason for the pedestrian simulation requirement, before the simulation phase itself, the modeling of pedestrians is a first and mandatory step. To accomplish this goal, a good knowledge of pedestrian behavior, both individual as well as collective (when in a crowd) is needed.

Pedestrians' movement in normal situation has its own dynamics. For instance, when walking along a sidewalk, in case of head-on encounters, a binary decision takes place: pedestrians need to choose whether to evade the other person on the right-hand or on the left-hand side. This decision process goes along with a significant decrease of walking speed (Moussaïd et al., 2009). This is an example of the interaction processes that happen between pedestrians walking at normal pace.

However, most of the normal behavior vanishes when pedestrians face an emergency situation. Similar effects can be observed for example in crowds trying to get the best seats at a concert or consumers running for sales (Almeida et al., 2011). Observations made for pedestrian crowds in emergency situations feature typically the same patterns. As people try to leave the building as fast as possible, the desired velocity augments which leads to some characteristic formations. As nervousness increases there is less concern about the comfort zone and finding the most convenient and shortest way. Such situations often happen during ingress and egress of public settings (Sime, 1995). The decision-making process of crowds and their dynamics under emergency situations is a crucial aspect that needs further investigation (Moussaïd et al., 2011).

Our methodology to study the dynamics of crowds, their interactions and both individual as well as collective behaviors encompasses the development of a microscopic pedestrian simulator using agent-based approach.

In order to conceive and devise such simulator, it is necessary to define crowd and determine its characteristics. Although there are several definitions for what characterizes a crowd, the broad meaning of the term can be defined by a series of features that define a crowd and its individuals (Challenger et al., 2009, p. 60), such as the ones listed:

- A considerable amount of people;
- People that come together in a specific area, that is, the same physical space;
- People that come together for a determined period of time, that is, not momentarily;
- Individuals in a crowd share objectives or have common interests;
- Individuals in a crowd show similar behaviors when acting in group;
- Individuals in a crowd interact with each other.

A crowd can also be characterized through a number of parameters, for instance, its size (number of individuals), density, period of time, importance of objectives, and interest sharing and level of adjustment to the environment (Challenger et al., 2009).

One paramount aspect to be considered in the development of the pedestrian simulator is to allow an easy configuration of the characteristics of the crowd to be simulated. For this reason it was envisaged the development of a tool that allows the creation of different agents and the definition of their characteristics by the designers.

8.2 Related Work

Research of pedestrians has been going for some decades from now, mostly based on direct observations, photographs, and videos (Sime, 1995; Helbing et al., 2002; Timmermans, 2009). Although attempts of modeling and simulating pedestrian movement exist for quite some time, this field of research recently received a clear boost in attention in a variety of disciplines, not only in the ones traditionally concerned with pedestrians such as transportation, urban planning and design, but also in applied physics, computer science, and artificial intelligence (Bouvier et al., 1997). In the latter case, pedestrian movement is often viewed as an interesting case to show properties of complexity theory and multi-agent models such as aggregate patterns emerging from simple principles applied to microscopic agents (Timmermans, 2009).

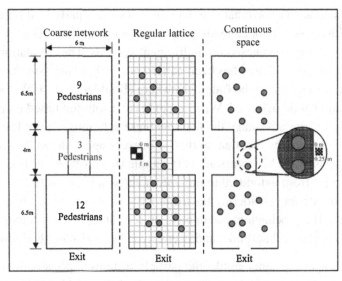

Figure 8.1: Illustration of the three more common representation models
Castle, 2007

The more common representation models (see Figure 8.1) use one of the three following formats (Castle, 2007):

1. Coarse Network (or physics-based models of particle flow) typical of macroscopic models;
2. Lattice or Cellular Automata;
3. Continuous space.

Although the first is intended for macroscopic simulation, the others are required for microscopic replication. One way of modeling space within the simulation context is by using the cellular automata concept, where models space is organized in a matrix with a two-dimensional array. The simulation technique uses a time-frame predefined in which the occupants can move from one position or node to another, assuming it is free or it is not an obstacle. Movement directions are limited to the eight possible nearby nodes. Another possibility is using the continuous space where movement is freely and each person can take whatever direction wants with no constraints.

Considering the entities interested in crowds and their behaviors, for instance, government agencies and insurance companies, a great effort has been dedicated to the study of this domain and several approaches have been proposed. Nowadays, there is a variety of different approaches and solution attempts at tackling the problem of efficiently simulate pedestrian behavior on emergency and evacuation scenarios. Throughout the various existing solutions, two groups stand out as the most explored.

The first one is based on solutions that simulate the crowd as a particle system, that is, considering the individuals as a homogeneous group in which everyone shares the same batch of behaviors. The second one is based on multi-agent systems. The simulation of crowds as particle systems is an approach that has been carried out with relative successful results. A few of these implementations can be found elsewhere (Bouvier et al., 1997) with more details on the implementation approach used. When it comes to a solution based on particle systems, we can consider as a major advantage the fact that these solutions require less processing capacity. This derives from the fact that, in this case, every agent shows the same behavior and, in most situations, these behaviors are very simple (Braun et al., 2003). Considering that human behavior varies from individual to individual and can be, sometimes, hard to predict, to simulate a crowd without considering the different behaviors between individuals will not be accurate enough. These behaviors can depend on a great amount of demographic characteristics, such as age, physical aptitude, and even psychological state (Kuligowski, 2009).

In the case of multi-agent systems, unlike the particle systems approach, the simulation will need a higher processing capacity due to the variety of agents and behaviors that are possible to simulate. Despite this disadvantage, this approach proves to be closer to reality because it allows a better representation of human behaviors (Castle, 2007).

Pedestrian simulators for studying the evacuation of buildings and urban areas include examples such as Exodus from Greenwich University, EVAC + FDS resulting of a collaboration between NIST in USA and VTT in Finland, as well as commercial applications like Massmotion, Legion, and Pathfinder (Kuligowski et al., 2010). Some authors, like Shao, propose a set of basic reactive heuristics for the pedestrian path planning process taking into account the interactions among agents and obstacles (Shao and Terzopoulos, 2005). Others use the Social Forces Model developed by Helbing and Molnar that uses a mathematical model to represent the attraction and repulse forces between pedestrians and the surrounding environment (Helbing and Molnar, 1995).

To address the need of a module that allows agent creation and configuration in the simulation environment, it is important to attend to the different approaches taken so far to tackle this same problem using BDI (Beliefs, Desires, Intentions) architectures. First of all, the reason for studying BDI agents is that over the past years, these agent architectures have been focused on a great amount of studies (Rao, 1996). Agent systems based on this architecture are applied to situations in which the agents are fed with continuous information in real-time and have to make decisions considering that information. The beliefs, desires and intentions are what drive the agent to make a decision, considering the information it has collected at that point. Generally, a predefined set of plans describing operators an agent can execute is used from which agents' instantiate a course of actions describing an intention.

As for practical ways to implement a BDI agent, AgentSpeak(L) is a logical language described in detail by (Rao, 1996) that allows for an easy definition of constructors defining the basic concepts of beliefs, desires and intentions. The implemented language can be interpreted as an abstraction of BDI systems that use the three main attitudes as data structures instead of modal operators. The implemented language allows the configuration of BDI agents through a simple syntax. The plan consists of a triggering event and a context. An example of the definition of a plan in AgentSpeak(L) is illustrated below.

```
+location(waste,X):location(robot,X) & location(bin,Y) <- pick(waste);
!location(robot,Y);
drop(waste).
```

The first successful attempt at using the AgentSpeak(L) language is described by (Machado and Bordini, 2002). The most common used interpreter for this language is Jason which is a java-based interpreter for an extended version of this language and is described in detail in (Bordini et al., 2007). Another approach worth mentioning is Jam, which is described as a BDI-based agent architectures and is explained in detail by (Huber, 1999). One of the problems of using Jam, which is developed in Java and UMPRS (a C++ version of Jam) is that for the user to define the agent configuration, it is required the user to have some experience with either language, namely Java or C++, respectively.

Considering that the objective of this work is to allow designers to better and easily describe the crowd to be simulated, the approach considered was to devise a heterogeneous multi-agent system. Toward achieving this goal the ModP (Esteves et al., 2009), a pedestrian simulator was used. For the syntax or tool to implement the configuration of the agents, since JAM and UMPRS are not viable options to be implemented because these architectures requires that the user has some experience with either Java or C++, which should not be required for the designers. The AgentSpeak(L) language was another possibility that was also considered. However, because of its complexity when compared with the agent architecture of the ModP simulator, the chosen approach was to create a simpler syntax that considers the needed agent parameters and plan structure.

8.3 ModP Pedestrian Simulator

ModP is a pedestrian simulator devised and currently under development within LIACC (Artificial Intelligence and Computer Science Laboratory), at University of Porto. This simulator is based on the multi-agent system paradigm and was conceived from scratch, in C++ with the Qt framework for interface, the OpenSteer library for steering behaviors and OpenGL for the 3-D viewer of the simulation. Its designation stands for **P**edestrian **Mod**eling (Esteves et al., 2009).

Initial goal was to create a tool for modeling pedestrians in multi-modal stations, allowing the optimization of transshipments while accounting for factors such as space, modes of transport, and pedestrian traffic flows, as well as the influence of all these parameters upon

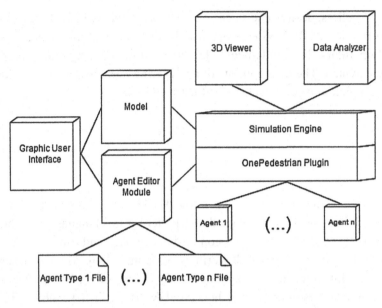

Figure 8.2: ModP Architecture with the Agent Editor Module

the decision making process of individuals using the station. Later, Aguiar adapted ModP to simulate and predict crowd behavior during emergency and evacuation situations (Aguiar et al., 2010). The simulator architecture has Graphic User Interface (GUI) allowing the model creation and editing as well as the Agent Editor Module, detailed in the following section. Other components are the 3-D viewer, the simulation engine, and a data analyzer (Figure 8.2).

Aguiar expanded the original ModP with the Agent Editor Module allowing the designers to simulate a broader range of crowds, creating the possibility to add new types of agents and configure the size of the crowd during the simulation. Also it was added the possibility of pre-defining a route for the agent (or agent type). For example, the agent has to withdraw money from the ATM machine, pick some documents (train tickets) in a specific place, and buy a cup of coffee, before leaving the building (suppose it was a train station).

The GUI editor is used for defining the layout of the building, the location of doors, source and drain points, that is, the places where pedestrians enter and exit the scenario (see Figure 8.3). The Points of Interest may also be defined, using the appropriate icons (ATM, coffee shop or train ticket) and finally the agent types.

8.4 Agent Editor Module

This new module allows the designers to create, configure, and deploy various agent types. It provides more realism to the simulation, as human beings vary in behavior and characteristics, as well as allows a precise configuration of the crowd, defined by the characteristics of

Figure 8.3: Snapshot of the ModP screen GUI

each agent type. In order to create an agent, the user must input the parameters that define that agent as well as its plan, that is, what are the objectives of this agent and the course of actions it is able to perform. The parameters that define an agent are its name, maximum steering force, maximum speed, comfort radius, mass, and vision range. In addition, each agent type must have a predefined plan. This plan is defined as a series of commands that the agent will follow orderly. The last command on the plan should always be the command *exit*, which represents the point where the evacuation ends, that is, the exit of the area or building where the agent is considered to be safe.

```
name: defaultAgent1
maxforce:300.4
maxspeed:1.5
radius:0.5
mass:1
visionrange:100
plan:find(10,-60);exit(10,-10)
```

In the snippet above it is possible to see all the agents' parameters and plan. In the next paragraphs, the existing commands will be explained in detail.

The plan of each agent type works as a list of actions that should be followed sequentially by the agent, that is, the agent can only start following the next command after completion of the previous one. In later developments, a timer should be added to each command in order to prevent impossible actions to obstruct the agent from successfully exit the building. Currently, there are only two commands that the user can add to the agent's plan. The main command is the *exit* command, which tells the agent where the exit of the building or area is located. The syntax for the exit command is *exit(x coordinate, y coordinate)*. The other is the *find* command that can be used in a wide variety of situations. It can symbolize that the agent must pick up some documents at a determined point in the building before exiting or that it should push an alarm button somewhere in the area. The syntax for this command is *find(x coordinate, y coordinate)*. The procedure to create a plan is very simple. A plan should always end with the *exit* command and all commands should be separated by a semicolon, as shown in the snippet below:

```
plan: find(60,-60);find(10,-60); exit(10,-10)
```

The execution of the plan is made using the Opensteer library. Both the *exit* and the *find* commands will use the *moveForSeek()* method. This method receives the coordinates of the objective as well as the parameters of the agent such as maximum speed, comfort radius, and so forth. and returns a force to be applied by the agent in order to move in the direction of the objective. This force is then aggregated with the forces returned from the *moveToAvoidObstacles()* method in order to prevent agents from colliding with each other and with obstacles throughout the simulation.

The *moveForSeek()* method still needs some improvement when it comes to path finding as the agents are only aware of an obstacle when it enters its comfort radius, resulting in some unintelligent routing behavior. In order to allow the users to configure which agents should enter each source and exit each drain, as well as in which proportions, a new tab was added to the drain and source properties dialog box in the graphic editor interface. This tab has a combo-box which displays all the available agents and an input box that allows the user to enter the percentage of agents of a certain type.

The changes made in the editor allow the dynamic definition of these parameters while the simulation is running, altering agents' behavior in real time. Changing the proportion of a type of agent will update the drain and source in the simulation as well.

8.5 Preliminary Results

Using the new functionalities described in the former section, a number of tests were carried out. The objective of these tests was to assert the influence of each parameter of an agent has on the simulation and what is the outcome of a change in one of those parameters or in the proportion of a determined agent type. For these tests, we have created three types of agents, namely child, adult, and elderly with the parameters shown in the Table 8.1 below.

The plan considered for each agent type was the same. Differences between agent types were on parameters' values. For instance, agents of the type child have half the mass of agents of the type adult, agents of the type adult have twice the maximum speed of agents of the type

Table 8.1: Agent types: child, adult, and elderly, and the correspondent parameters

agent name	maxforce	maxspeed	radius	mass	visionrange
child	250.0	1.0	1.0	0.5	75
adult	300.4	1.5	0.5	1.0	100
elderly	200.0	0.75	1.0	1.0	60

elderly. These values were defined only for test purposes. For achieving realistic results, demographic data would have to be collected in order to obtain sensible values to these parameters. The aim of these tests, as explained before, was to better show how different agents with different configurations or even their proportions in the simulation can alter the outcome of the simulation.

To test different agent types, some experiments were conducted featuring 30 agents distributed amongst the three created types (33% for child, 33% for adult and 34% for elderly). These agents were spawned at a specific point, 1,000 simulation units away from their objective. The aim of this test was to evaluate the changes in speed of each agent type when in a more realistic environment with other agents and different types of agents. The results arising from this test allow a better understanding of the influence agents of different types have on each other.

Analyzing the chart of Figure 8.4, it is possible to observe that there are differences on speed values among each agent type when interacting with each other. It was also noted that the speeds are different when comparing with agents of only one type in the same scenario. Most of the times, faster and more agile agents (in this case, agents of the type adult) can sometimes be slowed down by other agent types. Another interesting outcome one can observe

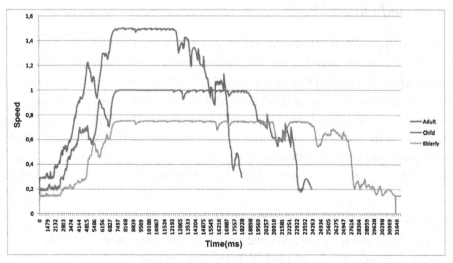

Figure 8.4: Speed over time of the three agent types (33, 33, and 34%)

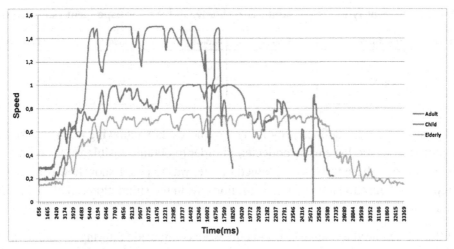

Figure 8.5: Speed over time of the three agent types (20, 40, and 40%)

in this chart is when agents come too close and start avoiding each other movement. For instance, between 20,300 ms and 21,600 ms it is possible to observe one of these situations, between agents of types child and elderly. When changing the proportions in the simulation to 40% for child, 40% for elderly (slower agents) and 20% for adults, the results also change as it can be seen on Figure 8.5.

As one can see in Figure 8.5, there is a considerable difference in the speed variation, when compared to the chart shown in Figure 8.4. It is possible to observe more and deeper variations of speed. This happens because of the greater amount of slow agents, impairing the movement of each other and also the faster agents.

8.6 Trends and Future Work

ModP was a first approach to develop an agent-based Pedestrian Simulation framework. Subsequently, further work showed that the development process from scratch is often hard and not worth the effort. Using COTS (Commercial Off-The Shelf) middleware or base software is sometimes a clever way to increase development and avoid expected problems, using this methodology as a shortcut to achieve faster and better results. Having this concept in mind, a new possibility is the use of the concept of Serious Games, using game engines to deal the computer graphics and animation details, releasing developers to focus on the domain issues rather than having to deal with all sorts of problems related with debugging and low level coding.

Some experiments were carried on based on the Unity3D game engine, resulting on ModP3D (Ribeiro et al., 2012) and EVA the EVAcuation Simulator (Silva et al., 2013), both works carried on as Master Dissertation works at FEUP, the Engineering Faculty of University of Porto.

Another research direction is using agent-based simulation frameworks such as NetLogo (Almeida et al., 2012), allowing rapid prototyping with interesting results.

The agent editor module that was added to simulator ModP, in order to make it more dynamic and allow the creation of more types of agents, needs further development to add more features and increase the tool practical utility for practitioners. For the agents' activities planning, more commands should be added to the syntax. Although the *exit* and *find* commands can suffice to describe some real situations, they are not enough to describe all the possible behaviors the agents should be able to have. Some examples of commands to be added are, for instance, the *follow* command to allow agents to chase some other agents. For instance, plans for child agents could include following adult agents and visitors at a museum could follow guides. Another important improvement to the agent editor module would be to add a timer to each command parameters in order to simulate the importance of that command. For example, instead of *find(x,y)*, the syntax would be *find(x,y,timer)*. Such timer could be interpreted as «the agent should only try to perform this action while it is safe to do so; when it becomes too dangerous to perform it (timer hits zero) the agent should proceed to the next action in the stack».

Another interesting enhancement to be implemented is a DXF importer for the environment model to be fed into the simulator. Some DXF interpreter libraries were studied, such as Kabeja and have shown to be too complex to convert the DXF data into the model structure in the ModP simulator. Nevertheless, this is an important issue that should be addressed in future works on this simulator, as it is of vital importance from the designers point of view to have some support in defining the environment area. Equally important, a Data Gathering module would allow, through different tools, the gathering of data related to pedestrian movement and behavior in order to refine the behavior of agents in the simulator. This module will be expected to gather information through two different approaches. The first one was to allow the introduction of avatars in the simulator. These avatars would be user-controlled individuals that would respond to a person's input in the simulation. The objective of this approach was to assert whether the routes taken by the agents were compliant with the routes taken by the user-controlled individual and with those results, refine the agents' path finding algorithm. The second approach was to use ubiquitous computing, for instance using RFIDs, to carry out real world evacuation simulations and collect the data from the RFID devices from each person in order to use that data to refine agents maximum speed and rotation force.

As mentioned earlier in the beginning of this section, development of in-house pedestrian simulator is a complex task, demanding many time and efforts that could be replaced by using other approaches. The Serious Games concept combined with agent-based simulation as well as using other means for pedestrian data collection, resulted in the conceptualization of a framework that was coined SPEED – Simulation of Pedestrian and Elicitation of

Emergent Dynamics. This is an undergoing research work at LIACC where one of the hot topics is using the possibility of using heterogeneous simulators connected through the HLA concept.

8.7 Conclusions

The importance of realistically simulating evacuation and emergency scenarios is the main focus of this project. The outcome of the addition of the proposed module to the simulator has proved to be very positive, as shown in Section 1.5, demonstrating that the differences between parameters and plans of the different types of agents presented in the simulation can alter significantly the speed and agility of each agent and, more importantly the evacuation time of a certain area. With this tool it is in fact possible for designers to establish a more realistic simulation of an environment. Using demographic data, it is therefore possible to use ModP to create and configure the crowd that will be present in the building or area, allowing sensitivity analysis for building managers or designers.

The added agent editor module increases the simulator's realism but there are many features to be added in order to use it as a reliable prediction tool. Two of the most important features to be included are the ones previously described, that is, import DXF files and create a data gathering module. Other improvements would be concerned to extending the kinematic library, OpenSteer. The path finding algorithms should be improved in order to obtain a more reliable simulation. The approach taken in this project can also be applied not only to evacuation and emergency situations but also to other pedestrian movement situations. For example, predicting visitors flow in a museum or in a subway station.

The development of pedestrian studies using different approaches was discussed in Section 1.6, pointing the use of game engines as an alternative to complete in-house development, thus saving time in development and debugging, as well as having the possibility of access to more advanced techniques with better results and less effort. This new development direction is being pursued by research teams at LIACC nowadays, having present, nevertheless, the aim which is to create and improve new techniques and tools for pedestrian modeling and simulation.

Acknowledgment

This work has been partially supported by FCT (Fundação para a Ciência e a Tecnologia), the Portuguese Agency for R&D, under grant SFRH/BD/72946/2010.

References

Aguiar, F., Rossetti, R., Oliveira, E., 2010. MAS-based Crowd Simulation Applied to Emergency and Evacuation Scenarios, in: IEEE ITSC2010 Workshop on Artificial Transportation Systems and Simulation (ATSS'2010). Madeira Island, Portugal.

Almeida, J.E., Rosseti, R., Coelho, A.L., 2011. Crowd simulation modeling applied to emergency and evacuation simulations using multi-agent systems. In: Sousa, A.A., Oliveira, E. (Eds.), DSIE'11-Sixth Doctoral Symposium on Informatics Engineering. Faculdade de Engenharia da Universidade do Porto. Engineering Faculty of Porto University, Porto, pp. 93–104.

Almeida, J.E., Kokkinogenis, Z., Rossetti, R.J.F., 2012. NetLogo Implementation of an Evacuation Scenario, in: WISA'2012 (Fourth Workshop on Intelligent Systems and Applications). Madrid, Spain.

Bordini, R., Wooldridge, M., Hübner, J., 2007. Programming Multi-Agent Systems in AgentSpeak using Jason (Wiley Series in Agent Technology). John Wiley & Sons. Chichester, UK.

Bouvier, E., Cohen, E., Najman, L., 1997. From crowd simulation to airbag deployment: particle systems, a new paradigm of simulation. J. Electron. Imaging 6, 94–107.

Castle, C.J.E., 2007. Guidelines for Assessing Pedestrian Evacuation Software Applications. Centre for Advanced Spatial Analysis.(UCL).

Challenger, R., C. W. Clegg, and M. A. Robinson. 2009. "Part 3: Literature on Crowd Behaviours." In Understanding Crowd Behaviour: Supporting Evidence, 60. The Cabinet Office Emergency Planning College. York, UK. ISBN: 978-1-874321-24-8

Cordeiro, E., Coelho, A.L., Rossetti, R.J.F., Almeida, J.E., 2011. Human behavior under fire situations – Portuguese Population. In: 2011 Fire and Evacuation Modeling Technical Conference. Baltimore, Maryland.

Esteves, E.F., Rossetti, R.J.F., Ferreira, P.A.F., Oliveira, E.C., 2009. Conceptualization and implementation of a microscopic pedestrian simulation platform. Proceedings of the 2009 ACM Symposium on Applied Computing – SAC '09 2105.

Helbing, D., Molnar, P., 1995. Social force model for pedestrian dynamics. Physical review E 51 (2), 4282.

Helbing, D., Farkas, I.J., Molnár, P., Vicsek, T., 2002. Simulation of pedestrian crowds in normal and evacuation situations. In: Schreckenberg, M., Sharma, S.D. (Eds.), Pedestrian and Evacuation Dynamics. Berlin, Springer, pp. 21–58.

Huber, M., 1999. JAM: A BDI-theoretic mobile agent architecture. Proceedings of the Third Annual Conference on Autonomous Agents, ACM Press, pp. 236-243.

Kuligowski, E.D., 2009. The Process of Human Behavior in Fires (NIST Technical Note 1632). National Institute of Standards And Technology.

Kuligowski, E.D., Peacock, R., Hoskins, B.L., 2010. A Review of Building Evacuation Models, second Ed. (NIST Technical Note 1680). National Institute of Standards And Technology.

Machado, R., Bordini, R., 2002. Running AgentSpeak (L) agents on SIM_AGENT. In: Meyer, J.-J., Tambe, M. (Eds.), Intelligent Agents VIII. Springer, Berlin, Heidelberg, pp. 158–174.

Moussaïd, M., Helbing, D., Garnier, S., Johansson, A., Combe, M., Theraulaz, G., et al. 2009. Experimental study of the behavioral mechanisms underlying self-organization in human crowds. Proceedings of the National Academy of Sciences. Biological sciences/The Royal Society 276, 2755–2762.

Moussaïd, M., Helbing, D., Theraulaz, G., 2011. How simple rules determine pedestrian behavior and crowd disasters. Proceedings of the National Academy of Sciences of the United States of America 108, 6884–6888.

Rao, A., 1996. AgentSpeak(L): BDI agents speak out in a logical computable language. Springer, Berlin Heidelberg, pp. 42-55.

Ribeiro, J., Almeida, J.E., Rossetti, R.J.F., Coelho, A., Coelho, A.L., 2012. Towards a serious games evacuation simulator. In: Troitzch, K.G., Möhring, M., Lotzmann, U. (Eds.), Twenty-sixth European Conference on Modeling and Simulation ECMS 2012. Germany, ECMS2012, Koblenz, pp. 697–702.

Shao, W., Terzopoulos, D., 2005. Environmental modeling for autonomous virtual pedestrians. In: SAE Symposium on Digital Human Modeling for Design and Engineering. Society of Automotive Engineers, 400 Commonwealth Dr, Warrendale, Iowa City, Iowa, USA.

Silva, José Fernando M., João Emílio Almeida, António Pereira, Rosaldo J. F. Rossetti, and António L. Coelho. 2013. "Preliminary Experiments with EVA - Serious Games Virtual Fire Drill Simulator." In 27th EUROPEAN Conference on Modelling and Simulation (ECMS 2013). Ålesund, Norway.

Sime, J., 1995. Crowd psychology and engineering. Safety Sci. 21, 1–14.

Timmermans, H., 2009. Pedestrian Behavior: Models, Data Collection and Applications, Traffic Safety. Emerald Group Publishing Limited, 27–43, ISBN 9-781-84855-750-5. Bingley, United Kingdom.

CHAPTER 9

Multi-Agent Active Collaboration Between Drivers and Assistance Systems

Jean-Paul A. Barthès, Philippe Bonnifait

Université de Technologie de Compiègne, UMR CNRS Heudiasyc, France

9.1 Introduction

Advanced Driver Assistance Systems (ADAS) are systems intended to help the driver in his driving activities. Technological solutions are many, like Adaptive Cruise Control (ACC), Intelligent Speed Adaptation (ISA) or Collision Warning Systems (CWS). When designed with a safe Human–Machine Interface (HMI), an ADAS should increase car safety and comfort.

Building a safe HMI requires careful attention as argued by ergonomics. For instance, Bruyas et al. (1998) have proposed guidelines to display information to the driver in running conditions. In this kind of problem, much effort is devoted to the choice of an efficient and safe strategy to display the information coming from the perception system of the vehicle or from cooperative perception, thanks to communication devices. Here, the driver is uniquely receiving information from the warning system. It is up to him to take into account the warning messages.

Some ADAS systems act in a completely different way, directly into the control inputs of the vehicle. Obstacle detection systems like the one presented by Broggi et al. (2002) can take the decision to brake, if the situation is estimated "very" dangerous. This kind of system is usually classified as "active safety". For some ergonomists, ADAS of this kind are called "dead driver systems" (Hoc and Debernard, 2002).

For several years, people focused on systems operating between these two extremes. Such approaches are sometimes referenced as "cognitive" (Althoff et al., 2007; Heide and Henning, 2006). Here, the problem is to devise a closer collaboration between the driver and the machine. Monitoring the driver's activities is the first key prerequisite (Murphy-Chutorian and Trivedi, 2010) but the problem goes far beyond. A collaboration can take place between the human and the machine to modify the setting of the parameters of the ADAS, for instance. A basic example is the problem of giving the destination address to the navigation system in order to compute a route. The collaboration can also occur during the operation, while

Advances in Artificial Transportation Systems and Simulation.
Copyright © 2015 Zhejiang University Press Co., Ltd. Published by Elsevier Inc. All rights reserved.

driving. This is typically the issue we consider is this chapter, by proposing to use a multi-agent system (called OMAS in the following) that acts as an interface between the driver and the ADAS, a real-time platform, PACPUS, managing sensors, collecting data from the vehicle, and implementing the ADAS function. By using speech recognition, the system presented here is able to understand sentences relative to the tuning of warning messages due to speeding in dangerous situations.

The remaining chapter is organized in five sections. We start by studying the classical role of multi-agent architectures for intelligent control in robotics. We present afterward a use case in which an ADAS system collaborates actively with the driver at a cognitive level. Then we detail the system that has been developed for this purpose. We report a critical review of the results that we obtained and we conclude by presenting future extensions of this research.

9.2 Multi-Agent Architecture for Intelligent Control

Multi-agent architectures provide an efficient framework to implement high-level, flexible, and modular control strategies, particularly for distributed and collaborative systems, for instance, the control of urban traffic in large cities (Balaji and Srinivasan, 2010). Some works have shown that they are also useful for autonomous systems. An example is the system used by (Beeson et al., 2008) for making autonomous a vehicle at the DARPA Urban Challenge. Here, a Multi-Agent System (MAS) is used for the environment perception and the choice of adequate vehicle's behavior. Several observers and several controllers run in parallel and are triggered by the MAS. Such a mechanism is called "reactive MAS" (Figure 9.1). Usually, such a high-level control gets along with a real-time, low-level architecture in charge of the feedback loops involving the actuators. Indeed, one has to make a distinction between the possibility of reasoning and low-level control. The reasoning system is rather slow with respect to the answer time needed for taking over the control of the car. When interacting

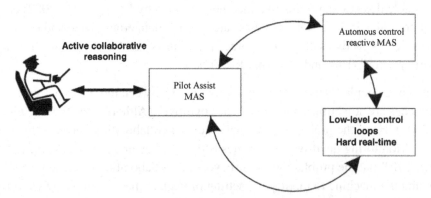

Figure 9.1: Possible use of Multi-Agent Systems (MAS) for automotive intelligent control (see box above/should be "autonomous").

with the driver, the cognitive task can be done in collaboration between the car and the driver. In this case, a third key component has to be considered explicitly. This is the "pilot assist" displayed in Figure 9.1. At this level, the reaction time is much longer since an active collaboration between the machine and the driver is necessary. Such an approach is somewhat similar to the situation in warships where the real-time defense system is automatic and fast, and the advising command system is much slower.

In the rest of this chapter, we focus on the active collaboration between the driver and an intelligent vehicle equipped with ADAS functions.

9.3 Use Case

Let us first introduce a use case: we are in a town (Compiègne) with a default speed limit of 50 km/h and want to drive up to "Lycée Charles de Gaulle," a particular high school. On the way, we encounter pedestrian crossings, bumps to slow the vehicle; we drive in front of the swimming pool before arriving at the high school. Figures 9.2–9.5 are screen captures from the experiment. In the figures you can see several windows: On the left, an obvious image showing the street, next to it a graph plotting the actual vehicle speed versus the maximal allowed speed, at the bottom right part of the interaction screen between the driver and his personal assistant named SUZY.

Figure 9.2 shows that the car is driving well under the speed limit. A first notch in the top profile indicates that the speed should be reduced (because of a pedestrian crossing area),

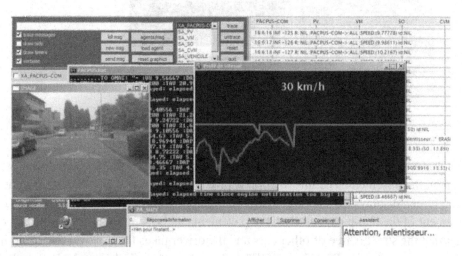

Figure 9.2: Alert due to bump just following a pedestrian crossing (the top left window is the OMAS multi-agent control panel, the top right window contains a graphics trace of the exchanged messages, the left window showing the road comes from the PACPUS system, the right window showing the graph is posted by Suzy, the bottom (tip of the iceberg) window is the Suzy interface window, the center background window is the Java trace of PACPUS).

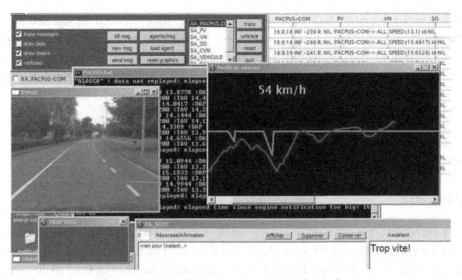

Figure 9.3: Speeding alert (bottom left window).

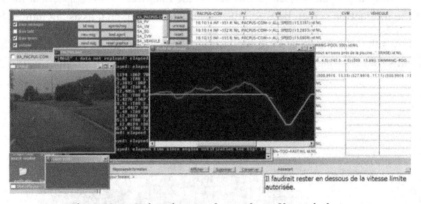

Figure 9.4: Swimming pool speed profile and alert.

immediately followed by a second notch indicating that the speed should be reduced because of a bump on the pavement. The information is provided by PACPUS. The bottom right part of the screen displays the message "Attention, ralentisseur… [1]" uttered by SUZY. The announcement of the bump painted onto the pavement can be seen in the street display on the left. After that the street is free of other cars and the driver takes the car over the speed limit. A message is uttered by SUZY: "Trop vite![2]" and a red window flashes at the bottom left of the screen. Suzy also emits a short beep.

---

[1] Watch the bump…
[2] Too fast!

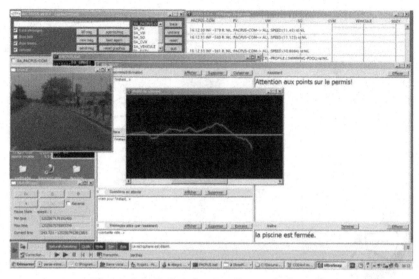

Figure 9.5: Swimming pool canceled profile.

The next figure shows that we have reached the swimming pool, and a maximum speed profile is posted showing that the speed should be reduced significantly when driving in front of the swimming pool at this particular time.

However, here the driver says "La piscine est fermée.[3]", which causes SUZY to remove the maximum speed profile and reset the maximum speed limit to 50 km/h (Figure 9.5).

Then the trip continues to the high school with similar situations. An important point is to notice that the driver can modify the behavior of SUZY by bringing new information to the system, for example, in the swimming pool area. While speeding the driver could tell SUZY something like "Je sais.[4]" which would cause SUZY to stop sending warnings to the driver until the next obstacle is detected. The driver can thus play an active role with respect to the ADAS.

In addition, SUZY is a regular personal assistant. SUZY can be connected to external sources and give information about the traffic or the weather or any information it can get from the Web, if additional agents are added to the system.

9.4 Architecture

The idea behind the current research is to allow a better communication between driver and vehicle. In order to have a high-level interaction, we need to have a vehicle equipped with an ADAS able to exploit information given by the driver.

---

[3] The swimming pool is closed.
[4] I know

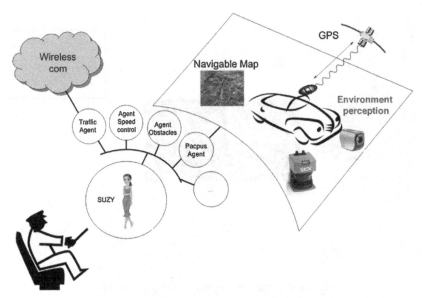

Figure 9.6: Multi-agent interface and organization.

The ADAS function has been prototyped using our PACPUS system[5]. It exploits mainly a precalculated itinerary, a navigable map, a positioning functionality, and an obstacle-detection system that fuses measurements coming from a Lidar (Light Detection and Ranging) and a stereo camera system to estimate a time to collision (Rodriguez Florez et al., 2010).

For developing the multi-agent system, we used previous work in which we developed "intelligent" agents for driving avatars to simulate interventions in dangerous plants (Edward et al., 2008) (by coupling a multi-agent platform with a virtual reality system), to propose a model where we associate a multi-agent system with a real-time platform on-board the vehicle.

Figure 9.6 presents the paradigm that we consider. All high-level information that has to be posted to the driver or that is coming from the driver has to pass through an agent called "SUZY". High-level information refers here to traffic information or ADAS alerts for instance. The agent called "PACPUS" is the one in charge of the ADAS.

9.4.1 The ADAS Function

The ADAS function considered in this work is an ISA function: the driver is informed of speeding if the vehicle is approaching a potentially dangerous area. This information can be addressed to the driver through an active gas pedal that for instance is made harder, a beep or

---

[5] http://www.hds.utc.fr

**Figure 9.7: Driver interface: active gas pedal, audio, and screen display.
The two cameras on the dashboard are used for eye-tracking.**

a display on a dedicated screen. Figure 9.7 illustrates the interface between the driver and the assistant that we consider in this work.

Two kinds of information are provided by the system. The first one is "static" and refers to what is called "Map- Enabled ADAS" meaning that the Digital Map is one of the principal inputs to the system and without it the function would not be possible. The information corresponds to points of interest (POI) that have been charted on the map, like stops, pedestrian crossings, or school entrances. To implement this method, we use an "Electronic Horizon" (EH) (Ress et al., 2005) that exploits a GPS positioning map, matched to a predicted path. The ADASIS v2 protocol (Durekovic and Smith, 2011) is a standardized Application Programming Interface (API) that can be used to implement a Map-Enabled ADAS. It is important to note that EH is much more reliable if the route has been planned before driving. The relevance of a POI alert can be modulated, thanks to calendar information. For instance, during vacation, schools are closed. This filtering issue is handled by the OMAS system.

The second kind of alert refers to "dynamic" events, like pedestrian crossing the road or traffic jams. This information is captured by the obstacle detection function based on Lidar combined with camera processing. Please note that we are not exploiting this information in time-critical situations, like many obstacle avoidance systems that already exist in some cars and apply a braking action as last resort. Here, this exteroceptive information is exploited as soon as possible in order to modulate the speed set-point of the ISA system.

Figure 9.8: The experimental car used to log the sensor data.

In our experiments, we have used a system called "D-BITE" (Bezet et al., 2006) to log the sensory information coming from the vehicle in the reported use cases. This system is able to collect a huge amount of time-stamped information (like different video streams). A player is then used to replay the data in a similar way to real-time conditions. This is an interesting functionality that is very useful for prototyping systems.

Figure 9.8 presents the vehicle that has been used in the experiments. The Lidar is located in the bumper and the stereo system is visible on the rooftop. The GPS antenna is installed on the rear.

9.4.2 OMAS

OMAS is a multi-agent platform intended to develop cognitive agents. OMAS agents are fairly complex. Each agent may have a number of different threads running in parallel. OMAS is described in detail by Barthès (2011), and the platform can be downloaded from www.utc.fr/~barthes/OMAS. In the experiment the agent platform is positioned between the user and the PACPUS platform as shown in Figure 9.6. The system contains six agents: PV, VM, CVM, SO, PACPUS-COM, and SUZY (Figure 9.9).

PV (Profils de Vitesse/Speed Profiles) contains a library of normalized speed profiles corresponding to various situations, for example, "pedestrian-crossing", "bump", "swimming pool", "school", "obstacle", and so forth. Receiving an ADAS message triggers the choice of the corresponding profile. The profile is made dimensional and sent to the VM agent (Figure 9.10).

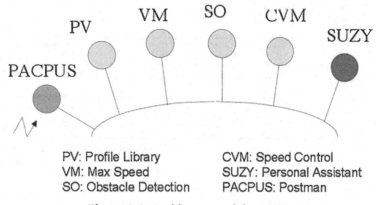

PV: Profile Library
VM: Max Speed
SO: Obstacle Detection

CVM: Speed Control
SUZY: Personal Assistant
PACPUS: Postman

Figure 9.9: Architecture of the MAS.

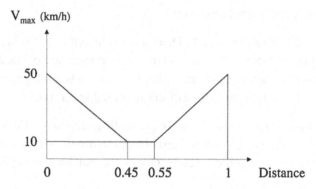

Figure 9.10: Maximal speed school profile.

VM (Vitesse Maximale/Maximal Speed) is an agent in charge of combining profiles to compute the maximal speed that the vehicle is allowed to reach on the itinerary. Every time a new profile is computed, it is sent to the CVM agent.

CVM (Contrôle de Vitesse Maximale/Maximal Speed Control) checks that the vehicle speed stays under the speed limit (with a small tolerance when exceeding it). When the vehicle is driving over the speed limit, a message is sent to SUZY, and SUZY normally warns the driver by telling something. We have seen that SUZY may be temporarily silenced.

All messages come from PACPUS-COM, a transfer agent (also called postman), interfacing the PACPUS real-time platform with the OMAS system. PACPUS-COM is in fact a gateway between the two systems that restructures the messages and sends them to the right agent with the proper format. PACPUS-COM may use point-to-point messages or broadcast messages

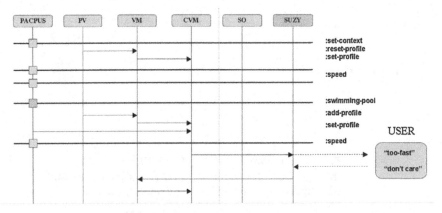

Figure 9.11: Agent lifelines (thicker lines represent broadcasts).

(e.g., for posting vehicle speed). In both cases, since we are using a UDP (User Datagram Protocol) protocol, we need a single message.

The SO (Surveillance d'Obstacle/Obstacle Detection) is linked to the vehicle obstacle detection system (Lidar and stereo-vision) and indicates if there are obstacles on the road or pedestrians walking on the sidewalk, for instance. If so, it sends a message to PV that will select a new profile to be merged with the current maximal speed profile.

SUZY is a personal assistant agent in charge of presenting information to the driver through the ADAS interface and decoding vocal requests from the driver. SUZY can also answer questions about stored data and could be connected to the outside world (Web).

9.4.3 How Does the System Work?

Figure 9.11 shows how the multi-agent system works in the context of the experiment. At first PACPUS broadcasts a message setting up the context: we are in town (thus speed is limited to 50 km/h). The PV agent computes an indefinite flat profile limiting the speed to 50 km/h and sends it to CVM that starts controlling that the received speed does not exceed the limit (following broadcasts). At some time PACPUS indicates that we are approaching the swimming pool (information obtained from GPS data). PM sends a library profile to VM that combines it with the current profile and ships it to CVM. CVM thus finds that the vehicle is driving too fast. It then sends a warning message to SUZY. SUZY posts the message, tells it to the driver and starts an alarm. However, the driver tells SUZY that he does not care or that the swimming pool is closed. SUZY sends a message to VM to remove the swimming-pool profile from the combined profile. This is done by VM that sends the new combined profile to CVM.

This experiment was intended to demonstrate that the OMAS platform could be coupled to the PACPUS real-time platform. Consequently, we do not see complex reasoning. However, each agent has a personal ontology containing domain concepts like a town, a road, a swimming pool, a school, or vacations. Each agent can contain rules or methods (including demons) to assess the situation and to decide an action. SUZY in particular should decide about a presentation policy, so that the driver is not overloaded with useless information. On the other hand, when a potential obstacle is detected, the system should decide whether or not to alert the driver, or, if there is not enough time to prepare the vehicle for a possible collision, by pretensioning seat belts, for example. This is possible because the PACPUS-COM agent is a gateway in both directions, meaning that messages can be sent to the PACPUS platform and action can be taken through this platform.

9.5 More technical data

This section gives some additional information for technically oriented people, although the system we developed is a proof of concept and is not meant to be an actual product.

9.5.1 Hardware

The current implementation of the system does not exploit any exteroceptive perception system (like cameras and Lidars). The ISA function uses a GPS receiver, a digital map, and speed measurement coming from the CAN (Control Area Network) bus of the vehicle. The POIs considered in this work have been added manually on the navigable map.

PACPUS and OMAS software runs on different machines installed on board. For convenience the D-BITE emulator replaying logged data, the ISA function and the OMAS platform were installed on the same notebook in order to simplify software development. The connection between the PACPUS platform and the OMAS platform uses an Ethernet cable. The vocal input is done through a microphone connected to the Multi-Agent System (MAS) notebook.

9.5.2 Software

The PACPUS platform is based on the D-BITE software that relies on an event-triggered real-time architecture. D-BITE is written in C++ and exploits a middleware called SCOOTR (Chaaban et al., 2005). SCOOTR is a fully distributed middleware based on a "client–server" approach. A hard real-time implementation has been developed with Real-Time Linux and Firewire interfaces between the computers. The system used in these experiments is soft real-time and has been implemented with MS Windows XP and Ethernet LAN.

The OMAS platform is written in Lisp (Allegro Common Lisp from Franz Lisp®). The vocal interface uses an old version of Dragon Naturally Speaking® from ScanSoft (today Nuance) and the interface was programmed using Visual Basic.

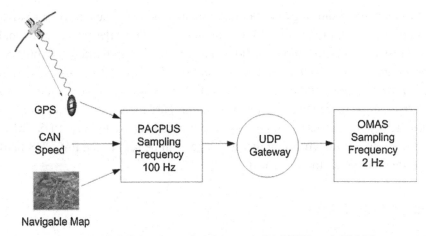

Figure 9.12. UDP connection between PACPUS and OMAS.

Transfers between the subsystems is done using UDP and implemented using Qt (QUdpSocket) (see Figure 9.12). A specific protocol between PACPUS and OMAS was designed with a very simple content language, for example,"= :VU 10 :DAP 83 :TAP 5.35." The typical fields to be used in the message are given in Table 9.1.

It is up to the PACPUS platform to package the data and to sequence messages at a rate that can be supported by the MAS (currently 2 Hz).

On the MAS side, we have three types of agents: the postman receiving messages from PACPUS and also capable of sending them to the vehicle (not done in this experiment), SUZY, a personal assistant, and service agents. All agents are on the same LAN (Local Area Network) loop (here they are in the same machine, but they could be distributed). Exchanges among agents are done by UDP broadcasts. The most interesting part is the communication between the driver and SUZY.

Table 9.1: Fields for the transfer content language.

Code	Meaning
DAP	Distance to a predefined target (e.g., school entrance)
VU	Vehicle speed
TAP	Time before arriving at predefined target
DV	Density of obstacles
DMV	Minimal distance to obstacles
TAV	Time to arrive at the obstacles
ARZ	Zone type, for example, school, swimming pool... (a symbol)

9.5.3 Communication with SUZY

The mechanism is based on a library of tasks that SUZY knows how to do and an ontology. When SUZY receives a new input (character string), she uses a library of tasks to determine which task is the most likely to be wanted. Based on a model of each task containing linguistic cues, the system ranks the tasks in decreasing values of a task score and removes tasks below a certain threshold. The first task is then executed, resulting in sending a message somewhere (specific agent or broadcast). SUZY has an ontology describing concepts and tasks. Table 9.2 describes the concept of a motorway that has two attributes: a speed limit (with a default of 130 km/h or 36.11 m/s) and a number identifying the motorway (e.g., A6).

Table 9.3 describes the task for killing constraints related to a swimming pool. Linguistic cues are defined as index patterns and for each cue a weight is given allowing us to compute a score for the corresponding task. A dialog reference is given that allows the triggering of a subdialog associated with that task.

Subdialogs are modeled as finite state machines (conversation graphs). In this case, the subdialog is very simple and has a single state shown (Table 9.4), where an answer is given to the driver and a message is sent to the VM agent. Dialogs may have any number of states and the system uses both linguistic patterns and the ontology to extract information from the user.

Subdialogs are modeled as finite state machines (conversation graphs). In this case, the subdialog is very simple and has a single state shown (Table 9.4), where an answer is given to

Table 9.2: SUZY's concept f motorway.

defconcept	(:en "motorway" :fr "autoroute")
:att	(:en "speed limit" :fr "vitesse limite")(:default 36.11)
:att	(:en "number" :fr "numéro") (:entry)

Table 9.3: SUZY's task for removing swimming pool constraints.

defindividual	"task"
:doc	:fr "Tâche d'enlèvement de la zone de la piscine"
"task name"	"remove swimming pool"
"performative"	:command
"dialog"	_remove-swimming-pool-conversation
"index pattern"	
	(:new "task-index" ("index" "piscine")("weight" .2)
	(:new "task-index" ("index" "fermée")("weight" .7)
	(:new "task-index" ("index" "pas ouverte")("weight" .7)
	(:new "task-index" ("index" "en grève")("weight" .7)

Table 9.4: SUZY's subdialog.

defstate	_rsp-entry-state
:label	"remove swimming pool dialog entry"
:entry-state	_remove-swimming-pool-conversation
:explanation	"user wants SUZY to stop sending warnings"
:text	"OK, j'ai bien noté"
:execute	send-message
	make-instance 'omas::message :type :inform
	:from :SUZY :to :VM
	:action :cancel-profile :args '(:swimming-pool)
:transition	:success

the driver and a message is sent to the VM agent. Dialogs may have any number of states and the system uses both linguistic patterns and the ontology to extract information from the user.

The number of tasks that one can have in the system is not limited and it is possible and not very difficult to add as many tasks as needed by the application. Some of the tasks can call web services if the vehicle has an Internet connection.

9.6 Discussion

This section discusses the limits of the experiment presented in the chapter. First, the system was implemented using a replay system of actual drives, able to replay all the data of the sensors like in real-time in the car. This has a significant advantage, namely the possibility to replay any part of the scenario as many times as necessary to test the system. The replay system can be seen on the bottom left of Figure 9.5 and a command console helps in verifying that the computer is not overloaded, which can occur when the replay speed is high.

9.6.1 Limitations of the Current Study

Since we have been working using emulation rather than with testing the system on-board, several features have been introduced without any error. For instance, locating pedestrian crossings and "bumps" on the road or reading speed limits using on-board cameras induces inevitably false alarms and miss-detections.

1. Locating Pedestrian Crossings and Bumps: The location of pedestrian crossings and bumps was manually coded in the EH since such information is not available in the current maps of Compiègne. It could be also obtained either by receiving signals from the environment (active road-side units), or by recognition of the corresponding road signs (see Figure 9.13). Currently, we have no road-side unit and no real-time program for analyzing the road signs. We think that a fusion of all these perception modalities should provide reliable information to our system.

Figure 9.13: Road sign for speed limit and bump.

2. Recognition of the Various Areas and of the Speed Limits: The recognition of the various areas like the town of Compiègne, the swimming pool location comes again from perfect data on the basis of global positioning and EH. In practice, recognition of speed limits should be more elaborate to take into account, for instance, road works. Moreover, extracting rough visual information from the picture shown in Figure 9.13 is not difficult with regard to the speed sign and bump sign. It is more difficult for the intermediate annotation (A 30 m) that adds information to the above sign. The information at times is a restriction on the application of the sign, for example, when the speed limit applies to trucks only. This should be provided either by having active road signs or by recognition by means of scene analysis.

3. Voice Recognition: There are also some issues concerning the voice recognition system. We have been using a rather old version of Dragon Naturally Speaking (version 7) in a quiet environment. The product has excellent performances in a quiet environment. It remains to be seen if this is still the case in a noisier environment. The new version of the software (version 11) sold currently by Nuance® could not be tested due to the unavailability of the corresponding SDK (programming interface). Voice interaction inside a car is a difficult problem regarding the vocal input (Shozakai et al., 1998). However, regarding output, the PACPUS platform can easily redirect the output to the car radio.

9.6.2 Advantages of the Proposed Approach

The main advantages of the proposed approach in addition to the vocal interface and two-way communications come from its modular architecture. Indeed, since the messages are sent in a broadcast mode, it is possible to add new features easily. Each agent will pick up the messages it needs for its particular reasoning. Some of the possible extensions are mentioned thereafter.

1. Choice of the Itinerary: This feature can be implemented easily through the API interface of the navigation system. The possible dialogs are then added to the library of dialogs of SUZY, corresponding to the various information/actions that can be obtained from the navigation system. The choice of the itinerary could be the result of a discussion, rather than a simple selection of choices provided by the GPS software.
2. State of the Car: In the same manner, dialogs concerning the condition of the car itself can be added to the library. It could be used by the driver to inquire about the state of the car or by the car to signal a specific problem. The information can be provided in real time.
3. New Sensors (for example, Eyes Tracking): The possibility of tracking the eyes of the driver, coupled with information about the speed of the car and the environment could be used to warn the driver in case of insufficient attention to the traffic.
4. New Sensors for Detecting Pedestrians: The advanced software developed by (Fayad and Cherfaoui, 2009) for detecting pedestrians (with confidence boxes as shown in Figure 9.14) coupled with dynamic information could be used to give more sophisticated warnings than when using only distance and density information.
5. Driver Profile: Driver profiles could be recorded and used to determine if the style of driving is within normal limits as given by the profile.

9.7 Conclusion and Future Developments

Trying to build a cognitive car can be attempted from two different perspectives: either one tries to introduce intelligence (meaning possibilities of reasoning) in the system controlling the car, or one can try to extend knowledge-based systems to introduce the necessary controls acting on the car. Both approaches are difficult and probably not so easy to develop. The approach we have chosen is to couple a system perfectly adapted to controlling the car, an ADAS implemented with PACPUS, to a system of hybrid agents that can be reactive, deliberative, and collaborative. The coupling is a loose coupling done at a fairly high level. Reasoning of symbolic data requires time and the real-time system must be slowed down, filtering data and transmitting them at a reasonable frequency (here 2 Hz).

We have developed a prototype that works in real-time with a data player that reads sensor measurements acquired by an experimental vehicle. This prototype is a proof of concept that opens up many perspectives.

We need now to develop our approach in several directions: (1) introducing more complex reasoning taking into account a model of the driver, a dynamic model of the environment, and some context; (2) integrating the system to the vehicle hardware (e.g., using the radio or

Figure 9.14: Pedestrian detected by a four-layer Lidar. Pedestrian detection confidence level (left bar) and Pedestrian recognition Confidence (right bar).

on-board hardware to improve communications); (3) integrating the prototype into a simulation system, which will allow us to run tests with different types of drivers in the laboratory rather than on the road; (4) adding more service agents to increase features, leading to richer dialogs. The first type of improvements would introduce driver profiles and driving patterns. Integrating the system to the vehicle hardware requires some work on the side of the PACPUS platform. Currently, the system is an addition to the vehicle and input, and output is directly handled by the OMAS notebook. It would be better to use the vehicle hardware. The third improvement is important, since finding drivers to run outside test drives is not so easy, and we have a full-size car simulator with additional hardware on the driver's side. Finally, the last type of improvement is not very difficult to do since the multi-agent system is open and one can add agents and services at any time.

Acknowledgment

This research has been conducted under the auspices of a project called "CODAVI" that regroups colleagues working in the same lab but in two different teams.

References

Althoff, M., Stursberg, O., Buss, M., 2007. Online verification of cognitive car decisions. 2007 IEEE Intelligent Vehicles Symposium.

Balaji, P., Srinivasan, D., 2010. Multi-Agent System in Urban Traffic Signal Control. Computational Intelligence Magazine, IEEE 5 (4), 43–51.

Barthès, J.-P.A., 2011. OMAS – a flexible multi-agent environment for CSCWD. Future Gener. Comp. Sy. 27, 78–87.

Beeson, P., O'Quin, J., Gillan, B., Nimmagadda, T., Ristroph, M., Li, D., Stone, P., et al. 2008. Multi-agent interactions in urban driving. JoPhA 2 (1), 15–29.

Bezet, O., Cherfaoui, V., Bonnifait, P., 2006. A system for driver behavioral indicators processing and archiving, in Ninth IEEE Conference on Intelligent Transportation Systems (ITSC 2006).

Broggi, A., Cerri, P., Ghidoni, S., Grisleri, P., Jung, H.G., et al. 2002. A new approach to urban pedestrian detection for automatic braking. IEEE Trans. Intell. Transport. Syst. 10 (4), 594–605, Dec, 2009.

Bruyas, M., Breton, B.L., Pauzié, A., 1998. Ergonomic guidelines for the design of pictorial information, Int. J. Ind. Ergonom. 21(5), 407–413. DOI:10.1016/S0169-8141(96)00081-9

Chaaban, K., Crubillé, P., Shawky, M., 2005. A distributed framework for real-time in-vehicle applications, IEEE Conference on Intelligent Transportation Systems (ITSC), Vienne, Autriche, 13–16.

Durekovic, S., Smith, N., 2011. Architectures of map-supported adas Intelligent Vehicles Symposium (IV), 2011. Germany, Baden-Baden, 5–9.

Edward, L., Lourdeaux, D., Lenne, D., Barthès, J-P.A., Burkhardt, J., 2008. Modeling autonomous virtual agent behaviors in a virtual environment for risk. IJVR 7 (3), 13–22.

Fayad, F., Cherfaoui, V., 2009. Object-level fusion and confidence management in a multi-sensor pedestrian tracking system, Lecture notes in Electrical Engineering, vol. 35.

Heide, A., Henning, K., 2006. The "cognitive car": a roadmap for research issues in the automotive sector. Ann. Rev. Control 30 (2), 197–203, [Online].Available: http://dx.doi.org/10.1016/j.arcontrol.2006.09.005.

Hoc, J., Debernard, S., 2002. Respective demands of task and function allocation on human-machine co-operation design: a psychological approach, Connection Science 14(4), 283–295. DOI:10.1080/0954009021000068745

Murphy-Chutorian, E., Trivedi, M.M., 2010. Head pose estimation and augmented reality tracking: an integrated system and evaluation for monitoring driver awareness. EEE Trans. Intell. Transport. Syst. 11 (2), 300–311.

Ress, C., Etemad, A., Hochkirchen, T., Kuck, D., 2005. Electronic Horizon–supporting ADAS applications with predictive map data. In ITS European Congress, Hannover, Germany-TS, vol. 18. 2005.

Rodriguez Florez, S., Fremont, V., Bonnifait, P., Cherfaoui, V., 2010. Time to collision estimation using a multimodal sensor fusion approach. San Diego (USA), IEEE Intelligent Vehicles Symposium, June 2010, 1-6.

Shozakai, M., Nakamura, S., Shikano, K., 1998. Robust speech recognition in car environments, in acoustics, speech and signal processing, 1998. Proceedings of the 1998 IEEE International Conference, 269–272.

CHAPTER 10

Computational Traffic Experiments Based on Artificial Transportation Systems: An Application of the ACP Approach

Fenghua Zhu, Zhengjiang Li

State Key Laboratory of Management and Control for Complex Systems, Institute of Automation, Chinese Academy of Sciences, Beijing, China

10.1 Introduction

The urban congestion problem is increasingly becoming a major issue in social, economic, and environmental concerns around the world. According to a recent survey, the 15 major cities in China are losing about one billion RMB (about $150,000,000) every day due to traffic congestions. The number of motor vehicles in Beijing, the capital of China, had exceeded 4,510,000 by September 12, 2010. So many vehicles have caused particularly serious congestion in this city. For example, the average time of Beijing residents' commuter trips has reached 52 min, which is the longest among all the cities in China (Niu, 2010).

The main difficulty of transportation modeling and analysis lies in the ability to reproduce an authentic transportation environment within the laboratory, as real world traffic scenarios are both too huge and too complex to be modeled (Wang et al., 2004). Traffic simulation has been considered as one promising tool in this area. Theoretically, simulation software can be used widely in transportation modeling and analysis. However, it still faces many challenges and its application is restricted to very limited areas. Conclusively, there are two insurmountable obstacles faced by the developers in the modeling and analysis road using simulation software. The first is how to generate individual travel demands for each person. Most traffic simulation software uses aggregating methods and requires historical origin-destination (OD) data as input. It is not only very costly to collect OD data in a wide area, but also very difficult, if not impossible, to transfer dynamic OD data into individual travel paths. Second, almost all simulation softwares focus on direct traffic-related activities alone and neglect other indirect facilities and activities, such as weather, legal, and social involvement. As environment exerts a profound influence over traffic, it is impossible to build an accurate model for transportation R&D using traditional methods (Wang, 2007).

Although the limitations of traffic simulation software were noticed soon after it was introduced into transportation study, there has been little done to deal with this for a long time.

Advances in Artificial Transportation Systems and Simulation.
Copyright © 2015 Zhejiang University Press Co., Ltd. Published by Elsevier Inc. All rights reserved.

However, the status has changed evidently since the early 2000s. First, the theory of traffic demand generation based on activity (TDGA) is becoming mature and has been applied in transportation planning in many developed countries (Bhat et al., 1999; Davidson et al., 2007; Arentze et al., 2011a; Roorda et al., 2009). In the United States, more than 40% of large metropolis plan organizations (MPOs) and 20% of the medium and small MPOs have adopted, or plan to adopt, TDGA models in their work (Arentze et al., 2011b). Second, the theory of artificial life and artificial society has proved to be a feasible approach in the research of the complexity of society and many achievements have been recorded. For example, Epstein and Axtel established "the world of sugar" to simulate the human society (Epstein et al., 1996), Los Alamos laboratory developed the epidemic simulation software based on individual behavior (Barrett et al., 2005), Research Triangle Institute (RTI) used and extended an iterative proportional fitting method to generate a synthesized, geospatially explicit, human agent database that represents the US population in the 50 states and the District of Columbia (Wheaton et al., 2009). All these achievements demonstrated that the integrative artificial society can be constructed from bottom up. Third, high-performance computing is becoming more and more popular. Not only the software and hardware of one computer have advanced rapidly, but also many networked computing technologies that can provide enormous computing capability utilizing the internet have been proposed, so that the heavy demand of computing and storage can be satisfied (Wang et al., 2007).

The ACP (Artificial Societies, Computational Experiments, and Parallel Execution) approach was originally proposed in (Wang et al., 2010a), as a coordinate research and systematic effort with those emerging methods and techniques, for the purpose of modeling, analysis and control of complex systems. Basically, this approach consists of three steps: modeling and representation with *A*rtificial societies; analysis and evaluation by *C*omputational experiments; and control and management through *P*arallel execution of real and artificial systems. The complex systems considered in the ACP approach usually have the following two essential characteristics (Wang et al., 2010a; Xiong et al., 2010):

- *Inseparability*. Intrinsically, with limited resources, the global behaviors of a complex system cannot be determined or explained by independent analysis of its component parts. Instead, the system as a whole determines how its parts behave.
- *Unpredictability*. Intrinsically, with limited resources, the global behaviors of a complex system cannot be determined or explained in advance on a large scope.

Clearly, real-world transportation systems, such as large-scale urban traffic systems, exhibit the two characteristics considered in the ACP approach. However, the motivation for employing the ACP here is mainly due to the lack of timeliness, flexibility, and effectiveness of the current simulation systems in transportation.

The focus of this chapter is to present our works and results of applying the ACP approach in modeling and analyzing a transportation system, especially establishing artificial transportation systems. The rest of this chapter is organized as follows: Section 10.2 introduces the process of growing artificial transportation systems from bottom up and lists some basic rules in the implementation; Section 10.3 proposes the method to model environment impact and demonstrates the process by modeling transportation scenarios in adverse weather; Section 10.4 shows the implementation architecture based on cloud computing; Section 10.5 verifies our method by illustrating one case study we carried out in Jinan, China; Section 10.6 draws conclusions with some remarks on future works and directions.

10.2 Growing Artificial Transportation Systems from Bottom Up

Transportation systems are becoming increasingly complex, nearly incorporating all aspects of our society. As more and more facilities and activities are involved in transportation, the connections between the transportation system and the urban environment are also getting closer and closer (Ahas et al., 2010; Hato, 2010). All these make the top–down reductionism method of traditional simulation very ineffective and there is still no effective method to model and analyze transportation systems. However, since one is inclined to be agreeable with simple objects or relationships, it is useful to build agent models on the basis of agreeable simple objects or relationships, then develop a bottom-up approach to "grow" artificial systems and observe their behaviors through interactions of simple but autonomous agents according to specified rules in given environments. In this context, the ACP approach is proposed to grow holistic artificial traffic systems (ATS) from bottom up (Wang, 2004; Miao et al., 2011; Zhu et al., 2011).

Here the main idea of ATS is to obtain a deeper insight of traffic-flow generation and evolution by modeling individual vehicles and local traffic behavior using basic rules and observing the complex phenomena that emerge from interactions between individuals. In the process of growing ACP-based ATS from bottom up, agent programming, and object-oriented techniques are extensively used for social and behavioral modeling (Wang, 2008). Figure 10.1 is the structure of one agent that represents one person in ATS. Though there still is not a universally accepted definition about an agent, it is widely accepted that an agent can be regarded as a computer system that is situated in some environment, and that is capable of autonomous action in this environment in order to meet its design objectives (Wang, 2005; Wang, 2002).

On the basis of the concepts and methods of artificial society and complex systems, ATS differs from other computer traffic simulation programs mainly in two aspects. First, the objective of traditional traffic simulation is to represent or approach the true state of actual traffic systems (Charypar et al., 2005), whereas the primary goal of ATS is to "grow" live traffic processes in a bottom-up fashion and provide alternative versions of actual traffic activities. In sociologist Theodor Adorno's words, ATS reveals traffic properties based on the belief that "only through what it is not will it disclose itself as it is." Second, ATS must deal with a wide range of information and activities. Most of the current traffic simulation focuses on

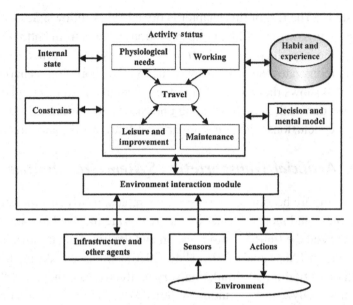

Figure 10.1: The structure of one agent in AT

direct traffic-related activities alone, whereas ATS generates their traffic processes from various indirect facilities and activities, such as the weather, and legal and social involvements. Some details about the first aspect will be explained in the following part of this section, whereas the second aspect will be the focus of the next section. Both of the introductions are very brief. For more details, readers can refer to Wang (2004); Miao et al. (2011); and Wang (2008).

Because individual's behaviors take the place of the OD matrix as input data in ATS, the first step in building ATS is to generate a reasonable population for a specified area. We implemented one separate module, namely an artificial population module (APM), for this task and, as mentioned before, modeled each person as one agent in this module. APM provides mechanisms to assign attributes to an agent as well as showing how these attributes change over time. In the design process of APM, plenty of theories and models in sociology and anthropology area are adopted. For example, the population structures in APM are divided into three types (Figure 10.2), namely, increasing type, decreasing type, and static type, which are also widely used by sociologists in classifying population.

Generally, travel is not undertaken for its own sake but rather to participate in an activity at a location that is separated from one's current location (Bhat, 1999). After constructing activity plans for each member of a population, travel demand can be derived from the fact that consecutive activities at different locations need to be connected by travel. Although one agent is carrying out its 24-h activity plan, its autonomy is mainly reflected in two aspects, one is his habit and experience, the other is the decision process that is based on his

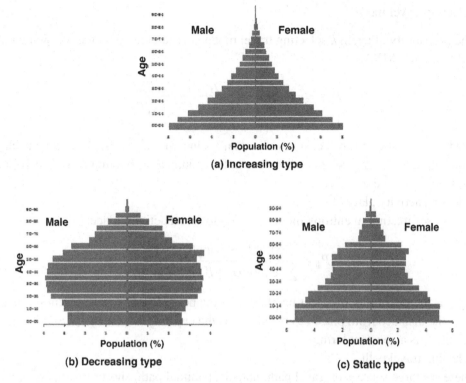

Figure 10.2: Three types of population structures

decision and mental model (Figure 10.1). All these features formed the foundation of our activity-based travel demand generation method, which fits well into the paradigm of multi-agent simulation and provides us with a feasible approach to generate an individual's travel demand.

Once the travel demand is generated, the agent needs to select the properties of the travel activity. There are mainly two types of factors that influence the selection process. One is the internal properties of the options, such as the capacity of place, the length of travel path, etc. The other relates to individual's psychology and behavior, such as his familiarity with the travel path, his feeling for the convenience of the travel mode, and so forth. Designing and establishing appropriate models for the latter is one of the focuses of our work. These factors are closely connected with the social and behavioral characteristics of individuals and their effects are usually expressed using natural language. Modeling the autonomous ability of one agent using discrete choice models (DCM) is a simple but compelling method that has been verified in various areas, especially in social and economic research, and is also adopted in our work. In the following, we will demonstrate some examples of them by showing their usages in the decision process of one agent.

- Selecting travel mode

  The probability of agent k selecting travel mode m is calculated by the following random utility model (RUM):

  $$P_m^k = \frac{\exp(e_k / M_{km} + f_k / T_{km} + g_k R_m)}{\sum_n \exp(e_n / M_{kn} + f_k / T_{kn} + g_k R_n)}$$

  where M_{km} is the ratio of travel cost to individual k's income, T_{km} is the travel time using mode m, R_m is the degree of convenience (a fuzzy indicator, ranging from 1 to 10) of mode m, e_k, f_k, and g_k are coefficients.

- Selecting activity place

  Agent uses maximum entropy model (MEM) to select activity place:

  $$P_{jli} = \frac{\exp(\alpha D_{ij} + \beta \log(C_j) + \gamma)}{\sum_r \exp(\alpha D_{ir} + \beta \log(C_r) + \gamma)}$$

  where P_{jli} is the possibility of selecting place j for the next activity when current place is i. D_{ij} is the distance from place i to place j, C_j is the capacity of place j, α, and β are coefficients, γ is a constant term.

- Selecting travel path

  There are three sources of travel path, namely, habitual path, shortest path, and minimum-cost path. Habitual path is a property of one agent, shortest path is a property of the road network and it is calculated in the initialization stage and kept constant until the road topology is modified, minimum cost path is system-wide dynamic information, which will be updated using real-time data and broadcast in the whole system at fixed intervals. The probability of agent k selecting path l is

  $$P_l^k = \frac{\exp(c_k / L_l + d_k F_{kl})}{\sum_t \exp(c_k / L_t + d_k F_{kt})}$$

  where L_l is the length of link l, Fk_l is the degree of the agent k's familiarity of path l. Fk_l is one fuzzy variable and the range of its value is 0~10. c_k, d_k are coefficients.

It is worth pointing out, besides providing feasible ways for modeling the decision process of one agent, there are many other advantages of modeling transportation systems from bottom up. For example, both Cyber-Physical Systems (CPS) and Cloud Computing are natural and embedded in this approach. As a matter of fact, CPS, as well as Cyber-Physical-Social Systems (CPSS) (Wang, 2010b), are special cases of intelligent spaces and an extension of our Intelligent Transportation Spaces (ITSp). Both were developed in our previous studies (Yang et al., 2007). As for cloud computing, it has already been used since the late 1990s in our work on agent-based control and management for networked traffic systems and other applications under the design principle of "Local Simple, Remote Complex" for high intelligence but low cost smart systems.

10.3 Modeling Environmental Impacts

It is well known that transportation is tightly connected with the social environment. From the microcosmic individual's psychology and driving behavior to macro-level travel and travel distribution, all are heavily impacted by the surrounding environment, such as economic, weather, and so forth (Hranac et al., 2006; Prevedouros et al., 2005; Ibrahim et al., 1994). The mechanisms by which the environment influences the traffic are very complex and there are still many disputes about how to represent them as a whole (Smith et al., 2004; Datla et al., 2008; Koetse et al., 2009). However, as to simple artificial objects, most of the current conclusions about the influences they received from the environment are consentaneous. So if simple objects and local behavior are modeled using these widely approved conclusions, the complex integrative phenomena that emerged are also expected to be understandable and agreeable. We call this principle "simple-is-consistent." Using this principle, we have established the rule bases to model the influences that a transportation subsystem received from other subsystems, as shown in Figure 10.3.

In the following, we will use adverse weather (rain) as an example to illustrate the models in ATS. For each individual in ATS, his experience of the influences of adverse weather can be

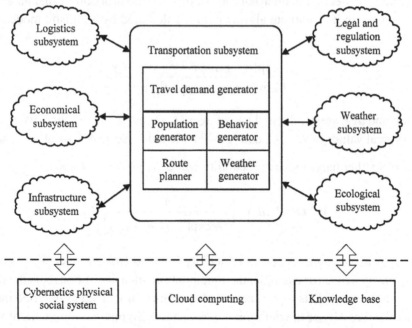

Figure 10.3: Environmental subsystems in ATS

expressed using a two-step model, which is composed of a travel demand generation process and traveling process. In the first step, the travel demand of individual i can be denoted as:

$$(\mathbf{A}_i, \mathbf{D}_i, \mathbf{P}_i, \mathbf{M}_i, \mathbf{ST}_i, \mathbf{ET}_i) =$$
$$(A_{i1}, A_{i2}, ..., A_{in}; D_{i1}, D_{i2}, ..., D_{in};$$
$$P_{i1}, P_{i2}, ..., P_{in}; M_{i1}, M_{i2}, ..., M_{in};$$
$$ST_{i1}, ST_{i2}, ..., ST_{in}; ET_{i1}, ET_{i2}, ..., ET_{in})$$

where $\mathbf{A}_i, \mathbf{D}_i, \mathbf{P}_i, \mathbf{M}_i, \mathbf{ST}_i, \mathbf{ET}_i$ are vectors of individual i's activities to be performed, travel desti-nations, travel paths, travel modes, start time, and end time, respectively. Usually, when ad-verse weather happens, an individual will adjust his activity plan to avoid unnecessary travel. According to the happening time of adverse weather, the adjustment measures include mov-ing up or putting off the happening time, lengthening or shortening the duration, and so on. If there has not been enough time in the schedule, some activities, especially those discretionary activities, such as shopping, sport, eating out, and entertainment, will be canceled eventually. Besides happening time and duration, destination and sequence can also be adjusted. All these adjustments can occur either before adverse weather, when we listen to the weather forecast and rearrange our activities accordingly, or in adverse weather, when adverse weather hap-pened unexpectedly. Obviously, no matter whenever the adjustment occurs, it can be well represented by adding new rules to our model.

Here we will use the probabilities of performing activity as an example to illustrate the influence of adverse weather. Under normal conditions, the probability that an agent i per-forms the activity k in its complete all-day plan is calculated by a logistic model, as shown below:

$$P_{ik} = \frac{\exp(\alpha_k \cdot gender_i + \beta_k \cdot age_i + \gamma_k)}{1 + \exp(\alpha_k \cdot gender_i + \beta_k \cdot age_i + \gamma_k)}$$

where $gender_i$ and age_i are gender and age of agent i, α_k and β_k are coefficients, γ_k is a con-stant term. Typical values of α_k, β_k, and γ_k will be listed in Section 10.4 of this chapter.

When adverse weather happens, the probability will be adjusted as follows:

$$W_{kj}^{\oplus}(P_{ik}, I_j, d_j) = \left(\frac{1}{1 + \exp\left[\delta_{kj}(I_j d_j - \phi_{kj})\right]} \right) P_{ik}$$

where δ_{kj} and ϕ_{kj} are constant properties of activity k and adverse weather W_j, I_j is the index to denote the intensity of W_j, for example, the precipitation intensity of rain, d_j is the duration of W_j. The term between brackets represents a S-shape function with an asymptotic maximum of one (either the intensity or the duration is zero) and an asymptotic minimum of zero (both the intensity and the duration are very high). δ_{kj} indicates the marginal effect of W_j at the inflexion point.

In the traveling process, adverse weather can influence an individual's driving behavior. Adverse weather can degrade the road performance due to the changes in the driving conditions (e.g., reduced visibility and pavement friction). As one consequence, it may cause a serious disturbance to the driver's reactions. All these can be represented by tuning an individual's driving parameters (e.g., free speed and free time headway). It is possible, then, to define and calibrate the actual functional relationship between these effects and changes in different parameters of driving models (Lam et al., 2008). Current driving models were mainly concerned with flow-based congestion effects and may not be applicable directly to the adverse weather conditions. To capture the rain effects, a new driving model is proposed on the basis of the conventional Intelligent Driver Model (IDM) (Treiber et al., 2000; Treiber et al., 2006), denoted as the Generalized Intelligent Driver Model (GIDM). The main idea of this model can be expressed using the following equation:

$$v'_i(I) = a \left[1 - \left(v_i / g(I, v_0^i) \right)^\delta \right. \\ \left. - \left(s_0 + h(I, T^i) v_i + \frac{v_i \Delta v_i}{2\sqrt{ab}} \right) \frac{1}{s_i} \right]$$

where $v'_i(I)$ is the acceleration of driver i in the next step when rainfall intensity is I. $v'_i(I)$ can be calculated using the following parameters:

- s_0, a and b are traffic jam distance, maximum acceleration and deceleration, respectively. The exponent δ is usually set to 4. These parameters are determined by transportation facilities and are usually the same for all drivers driving on the same road.
- v_i, s_i, and Δv_i are individual i's current speed, gap, and speed difference to the leading vehicle, respectively. These parameters represent current driving status of individual i.
- v_0^i and T^i are desired speed and safe time headway of individual i. The two parameters are determined by individual's features, such as psychology, age, and sex, and are specific for each driver.
- $g(I, v_0^i)$ and $h(I, T^i)$, which are the scaled functions of v_0^i and T^i, represent the adjustment of individual driving behavior in adverse weather.

Intuitively, the higher the rainfall intensity is, the lower the desired speed and the longer safe-time headway would be. We defined the two-scaled functions as follows:

$$g(I, v_0^i) = \frac{v_0^{max}}{1 + \left(\frac{v_0^{max}}{v_0^i} - 1 \right) \exp(pI)}$$

where v_0^{max} is the maximum individual's desired speed, and p is the coefficient that satisfies $p > 0$. We can see that, $g(I, v_0^i) \leq v_0^i$ is a decreasing function with respect to I implying that the driver's desired speed decreases while the rainfall intensity increases.

$$h(I, T^i) = \frac{T^{max}}{1 + \left(\dfrac{T^{max}}{T^i} - 1\right) \exp(-qI)}$$

where T^{max} is the maximum individual safe time headway, and q is the coefficient that satisfies $q > 0$. We also can see that, $h(I, T^i) \geq T^i$ is an increasing function with respect to I implying that the safe time headway increases whereas the rainfall intensity increases.

When I is 0, $g(I, v_0^i) = v_0^i$ and $h(I, T^i) = T^i$ implying that when there is no rain this new driving model is equivalent to the normal IDM model.

According to the functional form of GIDM, it can be seen that the higher the rainfall intensity the lower the acceleration. The GIDM model can be regarded as an extension of the normal IDM model. Under no rain condition ($g(I, v_0^i) = v_0^i$ and $h(I, T^i) = T^i$); the two models are equivalent.

It should be pointed out that, besides rain intensity, there are several other characteristics to represent rain, such as wind force, humidity, and visibility, as shown in our case study. For the sake of clarity, we use only rain intensity in designing GIDM. However, other characteristics can also be embedded in this model easily.

10.4 Implementation of Intelligent Traffic Clouds

Artificial transportation systems can use the autonomy, mobility, and adaptability of mobile agents to deal with dynamic traffic environment. However, the large-scale use of mobile agents will lead to the emergence of a complex, powerful organization layer that requires enormous computing and power resources. To deal with this problem, our implementation is based on intelligent traffic clouds (Youseff et al., 2008; Armbrust et al., 2009; Buyya et al., 2009).

Artificial transportation systems based on cloud computing have two roles: service provider and customer. All the service providers such as the test bed of typical traffic scenes, traffic strategy database, and traffic strategy agent database are all veiled in the systems' core: intelligent traffic clouds. The clouds' customers such as the urban-traffic management systems and traffic participants exist outside the cloud. Figure 10.4 gives an overview of artificial transportation systems based on cloud computing. The intelligent traffic clouds could provide traffic strategy agents and agent-distribution maps to the traffic management systems, traffic-strategy performance to the traffic-strategy developer, and the state of urban-traffic

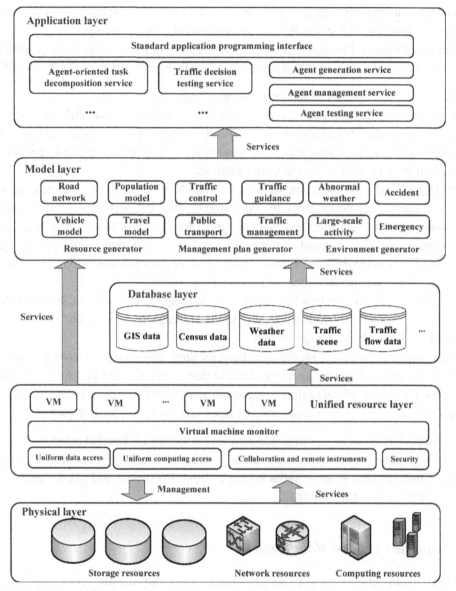

Figure 10.4: The architecture of artificial transportation systems based on cloud computing

transportation and the effect of traffic decisions to the traffic managers. It could also deal with different customers' requests for services such as storage service for traffic data and strategies, mobile traffic-strategy agents, and so on.

Using an intelligent traffic cloud, complex computing and massive data storage can be implemented on the cloud site, and the high performance can be achieved with low cost. All services are put into an intelligent traffic cloud. Besides transforming traffic control algorithms into control agents, the services also include agent performance test data and traffic detector data. Service consumers of an intelligent transport cloud include transport managers, control algorithm developers and transport control centers. According to the demands of service consumers, an intelligent transport cloud can provide the following services (Li et al., 2011):

- User identity authentication and permission management services.
- Transform services from public vehicle schedule algorithms to schedule agents. The services use a standard transform mechanism and universal API for traffic control algorithm developers. Once the schedule algorithm is implemented, the corresponding schedule agent can be generated in the cloud automatically.
- Performance test and evaluation services for vehicle schedule agent. On the basis of an artificial public transport system, the operation results, and performance of a vehicle schedule agent can be tested and evaluated in various traffic flows, using typical intersections and networks. The results, which are stored as a performance report of the agent, not only can provide decision support in selecting application conditions, but also can provide a reference for developers to improve algorithms.
- Storage management services for vehicle schedule agent. The services include vehicle schedule agent naming, redundancy, encryption, storage, and so on, and maintain load balancing of the whole storage system.
- Storage services for operation data and detector data. The services record the running process of vehicle schedule agents and traffic flow data collected by various detectors. By applying advanced data mining method in analyzing these data, traffic managers can acquire decision support to make and optimize vehicle schedule strategies.

With the support of cloud computing technologies, it will go far beyond other multi-agent traffic management systems, addressing issues such as infinite system scalability, an appropriate agent management scheme, reducing the upfront investment and risk for users, and minimizing the total cost of ownership.

10.5 Experiments and Validation

A field study on the effectiveness of ATS has been carried out in a district of Jinan city, the capital of Shandong Province, a populous region and a major economic power in northeast China.

We have focused on the area within the second ring of the Jinan urban traffic arterial network. This selected area, covering 255 km², east to Lishan Road, west to 12th Wei Road, south to 10th Jing Road, and north to Beiyuan Avenue, is the central business district of the city (see Figure 10.5). The area includes 410 sites that are directly related to traffic flow generation: 163 residential communities, 88 office buildings, 59 schools, 37 restaurants and hotels, 21 hospitals, 19 shopping malls, 13 recreational parks, and 10 sports facilities. An artificial

Figure 10.5: The Area of Field Study in Jinan for Computational Traffic Experiments

transportation system with 324 traffic nodes and 646 road links, called Jinan ATS, has been established for the selected area and various computational traffic experiments have been conducted on its basis.

This specific ATS provides us with a platform for conducting computational experiments for systematic, continuous application of computer simulation programs to analyze and predict behaviors of actual systems in Jinan in different situations. In the following, we will demonstrate how to model and analyze a transportation system based on Jinan ATS by showing the results of three computational experiments, which are constructing an activity plan for each individual, generating travel demand on the basis of activity and modeling the impacts of adverse weather.

10.5.1 Constructing Activity Plan for Each Individual

Travel demands are generated from an individual's activity plan, which serves as the foundation of ATS. Before carrying out computational experiments, the rationality of an individual's activity plan must be verified.

In ATS, we classified a person's activities into seven types: (1) work, (2) school, (3) hospital, (4) shopping, (5) sport, (6) eating (out), and (7) entertainment. Start time, end time and duration are three basic attributes for one specific activity. We suppose they all obey normal distribution and set their mean and standard deviations to different values. One shortcoming

Table 10.1: Attributes of activities (workday).

	Time Range (HH:MM)	Duration (min)	Probability		
			α_k	β_k	γ_k
School	[6:00–17:30]	$N(450, 20)$	0.1	0.01	12
Work	[6:30–20:00]	$N(480, 40)$	0.1	0.01	10
Hospital	[6:30–17:00]	$N(60, 10)$	−0.25	0.02	−1.65
Shopping	[10:00–20:30]	$N(90, 20)$	−0.91	0.01	0.56
Sport	[9:00–20:00]	$N(90, 10)$	0.13	0.02	−1.19
Eating	[16:00–19:00]	$N(60,10)$	0.25	0.01	−1.68
Entertain.	[15:00–20:00]	$N(90,10)$	0.57	0.03	−2.19

of normal distribution is its value range which is infinite, which may generate meaningless values, for instance, negative for start time. So we use bounded normal distribution (BND) instead of common normal distribution, as shown below:

$$\begin{cases} x \sim N(u,\sigma), & and \\ if \ \ x < u - 4\sigma & then \ \ x = u - 4\sigma, \ \ and \\ if \ \ x > u + 4\sigma & then \ \ x = u + 4\sigma \end{cases}$$

Calculated according to BND, the global attributes of these activities on workdays and weekends for Jinan ATS are listed in Table 10.1 and Table 10.2. Note that we use a time range to represent start time and end time here. Table 10.1 and Table 10.2 also listed the parameters for calculating the probabilities of activities, which have been explained in Section 10.3.

Based on the preconditions listed in Table 10.1 and Table 10.2, each individual will generate his specific travel demand using the discrete choice models in Section 10.2. And then the macro results will emerge naturally while numerous individuals are performing their activities. For example, Figure 10.6 presents the distributions of persons performing different activities from 5:00 PM to 11:00 PM in Jinan ATS, where population size in this area is set

Table 10.2: Attributes of activities (weekend).

	Time Range (HH:MM)	Duration (min)	Probability		
			α_k	β_k	γ_k
School	[6:00–17:30]	$N(450, 20)$	0.1	0.01	−2.4
Work	[6:30–20:00]	$N(480, 40)$	0.1	0.01	−2.2
Hospital	[6:30–17:00]	$N(320, 80)$	−0.18	0.02	−1.72
Shopping	[10:00–20:30]	$N(240, 60)$	−0.73	−0.01	0.64
Sport	[9:00–20:00]	$N(120, 40)$	0.13	0.02	−1.25
Eating	[9:00–19:00]	$N(90, 30)$	0.36	0.01	−1.81
Entertain.	[9:00–20:00]	$N(320, 80)$	0.63	0.03	−2.56

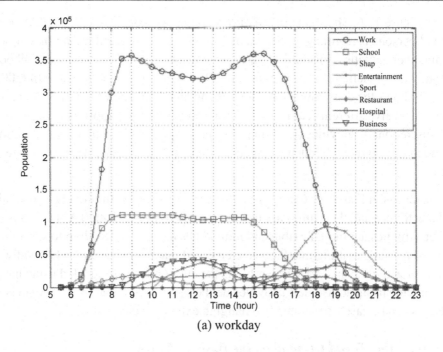

(a) workday

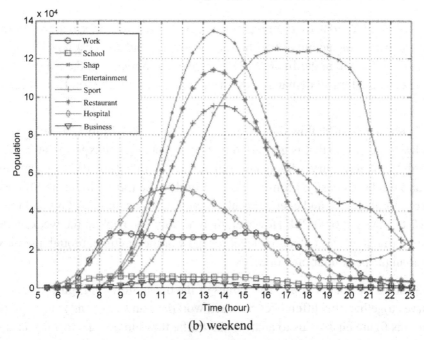

(b) weekend

Figure 10.6: Population Distributions performing different activities in a day

at 700,000. Figure 10.6a shows the distributions on one workday. We can see that the distributions of persons performing "work" and "school" are more regular than that of persons performing other activities. In addition, most people are performing "work" or "school" in the daytime and the population that participated in other activities is very small until 6:00 PM. Figure 10.6b shows the distributions at the weekend and, compared to Figure 10.6a, it possesses markedly different features. In Figure 10.6b, because more people are participating in discretionary activities (including shopping, sport, eating out, and entertainment), not only do these activities' frequencies increase sharply but also their time spans are extended.

Clearly, the results in Figure 10.6 are very consistent with reality. Intuitively, school and work are regular activities and their times are usually limited to between 8:00 AM and 6:00 PM, while other activities are more flexible and individuals have more freedom to schedule them. It is worth mentioning that Figure 10.6 is the emerged macro phenomena while individuals are doing their activity plans independently. As the environment is modeled using basic rules and each individual can adjust their activities deliberately, reasonable travel demands in various situations can be easily generated by changing experimental conditions.

10.5.2 Generating Travel Demand on the Basis of Activity

Just as travel is an induced activity in reality, an agent travels to perform activities in different places. After constructing activity plans both on workdays and at weekends for each person, travel demands in 1 week in the Jinan ATS can be derived directly. In the following, we will illustrate the generated travel demand with traffic flow data collected on Lishan Road in the Jinan ATS.

Figure 10.7 shows the traffic flow data of each day (from 5:00 AM to 12:00 PM) in 1 week (from Sunday to Saturday) that is generated by one computational experiment in the Jinan ATS. Though traffic flow curves fluctuate from day to day in Figure 10.7, we can distinguish workdays and weekends easily. From Monday to Friday, the traffic flow data follows an M-shape curve. Morning peak and evening peak are both obvious on workdays and their times are around 7:00 AM and 6:00 PM. However, there is only one peak in the weekend traffic flow and the traffic flow stays at a high level for most of the daytime, though the maximum is a bit lower than the peak hour on workdays.

Figure 10.8 shows the curves of average values of the traffic flows in Figure 10.7. Putting the two curves together, the differences between workdays and weekends are more obvious. In addition, this figure enables us to analyze the traffic flows in detail. On workdays, morning rush hours last from 6:00 AM to 8:00 AM, and the maximum flow in this period is about 500 vehicles per hour. Compared to the morning rush hour, the evening rush hour is shorter by about half an hour, and the maximum value is bigger by about 100 vehicles per hour.

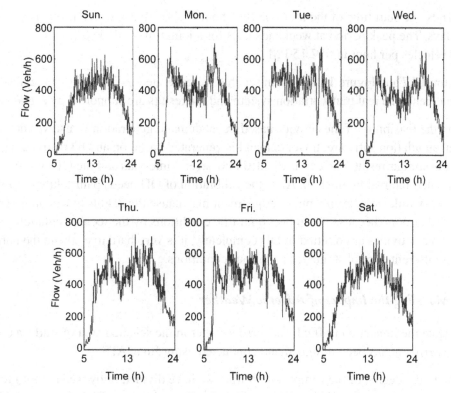

Figure 10.7: Traffic Flow from Sunday to Saturday

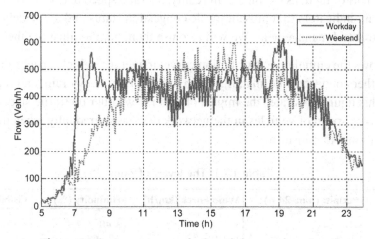

Figure 10.8: Average speed of workdays and weekends

At weekends, the start time of the morning rush hour is about 9:30, which is much later than on workdays. The peak period at weekends lasts for a long time and the traffic flow is higher than 500 vehicles per hour until 7:15 PM.

From Figure 10.7 and Figure 10.8, we can see that the generated traffic flows can represent the peak phenomena in a real transportation system and the results are supported by actual data.

Therefore, the feasibility of the activity-based travel demand generation is also verified. Using traditional simulation software, travel demand is generated based on an OD matrix and, to generate the similar result, one day is divided into several intervals and the OD matrixes have to be calibrated interval by interval. Also, the calibration of OD needs painstaking efforts. The process is very inflexible, even a minor adjustment may cause the results to be completely invalid. Furthermore, because the impact on traffic conditions by the local population's social and economic activities are omitted in OD completely, it is very hard to evaluate the performances of different control strategies in various conditions.

10.5.3 Modeling the Impact of Adverse Weather

To investigate the impact on traffic by adverse weather in the selected area of study, a computational experiment has been designed and conducted using Jinan ATS.

According to the degree of their impact on traffic, we have divided rainy weather into four levels, light rain, middle rain, heavy rain, and rainstorm, as shown in Table 10.3. Each kind of adverse weather is simulated for one whole day. Jinan ATS simulated the detailed traveling process of each individual in computational experiments and extensive evaluation indices can be generated. Because it is very difficult to show many of them due to space constraints, we will only show part of them as examples. In reality, average speed and vehicles in the network are two important indicators to represent traffic congestion status and are widely used in urban traffic control and management. We will also use them here to illustrate the results.

Figure 10.9 shows the cumulative distribution curve of average vehicle speeds on one day under five weather conditions: normal, light rain, middle rain, heavy rain, and rainstorm, respectively. This figure illustrates the impact of adverse weather on traffic status. As expected, the average speed of vehicles in the network decreases gradually when the weather changes from normal to rainstorm.

Table 10.3: The levels of rain.

	Precipit. (mm/24 h)	Wind Force (km/h)	Humidity (%)	Visibility (m)
Light rain	<10	<5	<40	>200
Middle rain	10~25	6–19	30~60	100~200
Heavy rain	25~100	20~38	50~80	20~100
Rainstorm	>100	>38	>80	<20

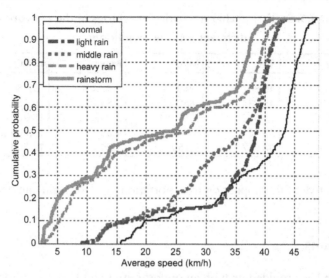

Figure 10.9 The Cumulative Distribution Curves of Average Vehicle Speeds under Different Weather Conditions

Table 10.4 shows some statistical characteristics of average vehicle speeds under different weather conditions. We can see that when weather conditions are getting worse, the mean, minimum and maximum are almost all decreased (some exceptions may be caused by random errors), while the standard deviation is increased. The 15% and 85% quartiles, which are common indices used in urban traffic evaluation, are also shown in Table 10.5.

Figure 10.10 shows the cumulative distribution curves of vehicles in the network on the same day as Figure 10.9 under different weather conditions. As expected, the number of vehicles in the network increases gradually when the weather changes from normal to rainstorm. Table 10.4 also shows some statistical characteristics of vehicles in the network under different weather conditions. We can see that when the weather condition is getting worse, almost all these indices are increased (some exceptions may be caused by random errors), which means the traffic status is getting worse and worse. One interesting conclusion that can be drawn from Figure 10.8 and Figure 10.9 is that the impact of light rain and middle rain is very close, while the impact of heavy rain and rainstorms is also very close. It seems

Table 10.4: Statistical characteristics of average vehicle speeds under different weather conditions.

	Mean	Std.	Min.	Max.	15%Q	85%Q
Normal	38.73	9.23	15.66	48.85	26.01	46.05
L. rain	34.87	8.89	9.14	43.82	24.53	40.89
M. rain	32.46	9.29	10.94	44.10	21.22	40.77
H. rain	23.77	14.08	2.61	43.82	6.38	39.43
RS	21.60	13.75	2.32	45.91	4.33	37.16

Table 10.5: Statistical characteristics of vehicles in the network under different weather conditions.

	Mean	Std.	Min.	Max.	15%Q	85% Q
Normal	35,079	40,847	950	1,77,150	9,250	52,000
L. rain	47,226	45,734	850	1,89,950	10,350	76,900
M. rain	53,293	52,512	1,150	2,14,000	10,400	1,08,600
H. rain	77,581	57,661	1,000	1,85,850	17,700	1,46,600
RS	85,058	62,067	1,150	1,90,250	19,300	1,63,500

that, in the perspective of the impact on transportation systems, rain can be represented using even fewer categories.

10.6 Conclusions

The ACP approach has provided us with an opportunity to look into new methods in addressing transportation problems from new perspectives. In this chapter, we present our works and results of applying the ACP approach in modeling and analyzing transportation systems, especially carrying out computational experiments based on artificial transportation systems. Two aspects in the modeling process are analyzed. The first is growing an artificial transportation system from bottom up using agent-based technologies. The second is modeling environmental impacts on the simple-is-consistent principle. Three computational experiments are carried out in one specific ATS, Jinan ATS, in the case study and numerical results are presented to illustrate the applications of our method.

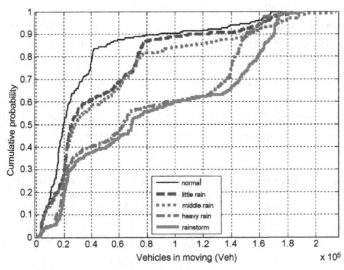

Figure 10.10: The cumulative distribution curves of vehicles in the network under different weather conditions

Recently, an intensified effort has been launched to set up standards and procedures to construct artificial transportation systems based on the ACP approach. Unlike conventional traffic simulation programs, those ATS are intended to be running continuously in cyberspace through web computing and computer gaming technologies, just like real traffic systems in real cities. We believe it has opened up a new field in a new direction that could significantly advance the level of effectiveness and intelligence of intelligent transportation systems and promote their future applications.

Acknowledgment

The authors would like to express their sincere thanks to Prof. F.-Y. Wang and all other colleagues and students in the Laboratory of Complex Adaptive Systems for Transportation (CAST), Institute of Automation, Chinese Academy of Sciences.

References

Ahas, R., Aasa, A., Silm, S., Tiru, M., 2010. Daily rhythms of suburban commuters' movements in the Tallinn metropolitan area: Case study with mobile positioning data. Transport. Res. C-Emer. 18 (1), 45–54.

Arentze, T., Ettema, D., Timmermans, H.J.P., 2011a. Estimating a model of dynamic activity generation based on one-day observations: Method and results. Transport. Res. B-Meth. 45 (2), 447–460.

Arentze, T., Timmermans, H.J.P., 2011b. A dynamic model of time-budget and activity generation: Development and empirical derivation. Transport. Res. C-Emer. 19 (2), 242–253.

Armbrust, M., Fox, A., Griffith, R., Joseph, A.D., Katz, R.H., 2009. Above the clouds: A Berkeley view of cloud computing. EECS Department, University of California, Berkeley, Tech. Rep., UCB/EECS-2009-28.

Barrett, C., Eubank, S., Smith, J., 2005. If smallpox strikes Portland. Sci. Am. 292 (3), 54–61.

Bhat, C.R., Frank, S.K., 1999. Activity-Based Modeling of Travel Demand. Springer.

Buyya, R., Yeoa, C.S., Venugopala, S., Broberga, J., Brandicc, I., 2009. Cloud computing and emerging IT platforms: vision, hype, and reality for delivering computing as the 5th utility. Future Gener. Comp. Sys. 25 (6), 599–616.

Charypar, D., Nagel, K., 2005. Generating complete all-day activity plans with genetic algorithms. Transportation 32 (4), 369–397.

Datla, S., Sharma, S., 2008. Impact of cold and snow on temporal and spatial variations of highway traffic volumes. J. Transport Geogr. 16 (5), 358–372.

Davidson, W., Donnellyb, R., Vovsha, P., Freedmanc, J., Rueggd, S., Hickse, J., Castiglionef, J., Picadog, R., 2007. Synthesis of first practices and operational research approaches in activity-based travel demand modeling. Transport. Res. A-Pol. 41, 464–488.

Epstein, J.M., Axtell, R.L., 1996. Growing artificial societies: Social science from the bottom up[M]. Mit Pr.

Hato, E., 2010. Development of behavioral context addressable loggers in the shell for travel-activity analysis. Transport. Res. C-Emer. 18 (1), 55–67.

Hranac, R., Sterzin, E., Krechmer, D., Rakha, H., Farzaneh, M., 2006. Empirical Studies on Traffic Flow in Inclement Weather. In: Administration, F.H. (Ed.), Publication No. FHWA-HOP-07-073, Washington, DC.

Ibrahim, A., Hall, F.L., 1994. Effect of adverse weather conditions on speed-flow-occupancy relationships. TRB (1457), 184–191.

Koetse, M., Rietveld, P., 2009. The impact of climate change and weather on transport: An overview of empirical findings. Transport. Res. D-Tr. E. 14, 205–221.

Lam, W., Shaoa, H., Sumaleea, A., 2008. Modeling impacts of adverse weather conditions on a road network with uncertainties in demand and supply. Transport. Res. B-Meth. 42 (10), 890–910.

Li, Z., Chen, C., Wang, K., 2011. Cloud Computing for Agent-Based Urban Transportation Systems. IEEE Intell. Syst. 26 (1), 73–79.

Miao, Q., Zhu, F., Lv, Y., Cheng, C., 2011. A Game-Engine-Based Platform for Modeling and Computing of Artificial Transportation Systems. IEEE Trans. Intell. Transport. Syst. 12 (2), 343–435.

Niu, W., 2010. 2010 report of the new urbanization of China. Science Press, Beijing, China (in Chinese).

Prevedouros, P., Chang, K., 2005. Potential Effects of Wet Conditions on Signalized Intersection LOS. J. Transport. Eng. 131 (12), 898–903.

Roorda, M., Carrasco, J.A., Miller, E.J., 2009. An integrated model of vehicle transactions, activity scheduling and mode choice. Transport. Res. B-Meth. 43 (2), 217–229.

Smith, B. L., Byrne, K.G., Copperman, R.B., Hennessy, S.M., Goodall, N.J., 2004. An Investigation into the Impact of Rainfall on Freeway Traffic Flow. Transportation Research Board, Washington, D.C.

Treiber, M., Hennecke, A., Helbing, D., 2000. Congested traffic states in empirical observations and microscopic simulations. Physical Review E 62, 1805–1824.

Treiber, M., Kesting, A., Helbing, D., 2006. Delays, inaccuracies and anticipation in microscopic traffic models. Physica A (360), 71–88.

Wang, F., 2002. On Some Basic Issues in Network-Based Direct Control Systems. Acta Automatica Sinica 28 (Supp 1), 171–176, (in Chinese).

Wang, F., Tang, S., 2004a. Artificial societies for integrated and sustainable development of metropolitan systems. IEEE Intell. Syst. 19(4), 82–87.

Wang, F., 2004b. Concept and Framework of Artificial Transportation System. J. Complex Systems Complexity Sci. 1, 52–57.

Wang, F., 2005. Agent-based control for networked traffic management systems. IEEE Intell. Syst. 20 (5), 92–96.

Wang, F., Carley, K.M., Zeng, D., Mao, W., 2007. Social computing: from social informatics to social intelligence. IEEE Intell. Syst. 22 (2), 79–83.

Wang, F., 2007. Toward a paradigm shift in social computing: the ACP approach. IEEE Intell. Syst. 22 (5), 65–67.

Wang, F., 2008. Toward a revolution in transportation operations: AI for complex systems. IEEE Intell. Syst. 23 (6), 8–13.

Wang, F., 2010a. Parallel control and management for intelligent transportation systems: concepts, architectures, and applications. IEEE Trans. Intell. Transp. Syst. 11 (3), 630–638.

Wang, F., 2010b. The emergence of intelligent enterprises: from CPS to CPSS. IEEE Intell. Syst. 25 (4), 107–110.

Wheaton, W.D., Cajka, J.C., Chasteen, B.M., Wagener, D.K., 2009. Synthesized Population Databases: A US Geospatial Database for Agent-Based Models. RTI Press publication No. MR-0010-0905. NC: RTI International.

Xiong, G., Wang, K., Zhu, F., Chen, C., An, X., Xie, Z., 2010. Parallel traffic management for the 2010 Asian Games. IEEE Intell. Syst. 25 (3), 81–85, 2010.

Yang, L., Wang, F., 2007. Driving into intelligent spaces with pervasive communications. IEEE Intell. Syst. 22 (1), 12–15.

Youseff, L., Butrico, M., Da Silva, D., 2008. Toward a unified ontology of cloud computing, IEEE Grid Computing Environments Workshop. GCE', 08.

Zhu, F., Li, G., Li, Z., Chen, C., Wen, D., 2011. A case study of evaluating traffic signal control systems using computational experiments. IEEE Trans. Intell. Transport. Syst. 12 (4), 1220–1226.

CHAPTER 11

Implementation of Transit Signal Priority and Predictive Priority Strategies in ASC/3 Software-in-the-Loop Simulation

Milan Zlatkovic*, Peter T. Martin**, Ivana Tasic†

*Department of Civil and Environmental Engineering, University of Utah, Salt Lake City, Utah, USA;
**Department of Civil Engineering, New Mexico State University, Las Cruces, New Mexico, USA;
†Department of Civil and Environmental Engineering, University of Utah, Salt Lake City, Utah, USA

11.1 Introduction

Microsimulation software packages are successfully applied to all types of traffic signal control simulation. Implementation of traffic control logics in traffic microsimulation provides modeling of both pretimed and actuated traffic control. In most traffic microsimulation packages the traffic control system is emulated within the software. This is called Emulator-in-the-Loop (EIL), because this emulator does not have any counterpart in the field. EIL can also be achieved through the Vehicle Actuated Programming (VAP) interface. In this case, the traffic control mechanism is developed in a programming language (Visual Basic, C++, Java, and alike) and is called through the microsimulation interface. VAP allows for a more customized traffic control than the built-in EIL controllers can offer.

Emulated control is later replaced with the real traffic control hardware. One or more signal controllers are integrated with the traffic microsimulation software. This enhancement of communication between the traffic simulator and traffic controller requires that an actual hardware controller be driven by the simulation through a process called Hardware-in-the-Loop simulation (HIL) (Urbanik and Venglar, 1995).

The most advanced form of traffic simulator and traffic controller interface is Software-in-the-Loop simulation (SIL) (Urbanik et al., 2006). The SIL concept allows the simulation of several virtual controllers under simulation software without the cost and complexity of physical controllers and controller-interface devices. SIL can also run in a mode that is faster than real time, facilitating simpler and less time-consuming simulation runs, something that the HIL concept cannot provide.

Advances in Artificial Transportation Systems and Simulation.
Copyright © 2015 Zhejiang University Press Co., Ltd. Published by Elsevier Inc. All rights reserved.

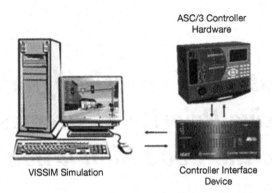

Figure 11.1: ASC/3 – VISSIM HIL concept

11.1.1 HIL Concept

In the HIL concept, the data generated from the simulation model vehicle detectors are first sent to the controller interface device (CID) (Urbanik et al., 2006; Stevanovic et al., 2009). The CID provides the interface between the computer that is running the traffic microsimulation and the discrete logic levels of the control pins in the traffic controller. After receiving the data through the CID, the traffic controller analyzes the input, determines the status of the signal control according to its control logic, and sends the data about the signal control status back to the simulation model through the CID. During every simulation time step, the data exchange is conducted between the simulation model, the CID, and the traffic controller. The CID functions as a bridge between the electrical signals of the computer and those of the traffic signal controller. The real traffic controller determines the status of traffic signals through the CID integration, replacing the internal control logic emulated in the simulation software. Figure 11.1 shows the HIL concept of the Econolite's Advanced System Controller series 3 (ASC/3)-traffic controller and VISSIM microsimulation.

11.1.2 SIL Concept

The SIL concept was developed to overcome the major HIL problems related to the complexity of physical controllers and CID devices. The main idea of the SIL is that both the simulation program and virtual traffic controller are running on the same computer, with an interface that allows communication between them.

Two well-known SIL applications have been developed in recent years: Siemens's NextPhase, which is linked to CORSIM and VISSIM, and ASC/3 which connects to VISSIM (Urbanik et al., 2006; Stevanovic et al., 2009). PTV America and Econolite Control Products, in co-operation with the University of Idaho (the MOST Project), have developed an ASC/3 SIL controller embedded in VISSIM (Urbanik et al., 2006).

Several virtual ASC/3 controllers can be integrated with VISSIM. These controllers are compliant with the National Transportation Communications for Intelligent Transportation Systems

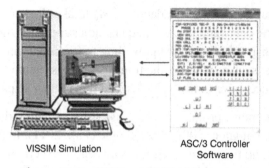

VISSIM Simulation

ASC/3 Controller
Software

Figure 11.2: ASC/3 – VISSIM SIL concept

Protocol (NTCIP) and operate from the same code base as the ASC/3 hardware controllers, making them nearly identical. This is a big advantage over emulators and custom-developed VAP simulation traffic controllers, because all the features and options are the same in both versions.

However, SIL does not have the features of a real controller that supports the communications within a cabinet or centralized traffic signal system. This is the major disadvantage of the SIL concept. The ASC/3 – VISSIM SIL concept is given in Figure 11.2.

11.1.3 Priority Strategies

Transit Signal Priority (TSP) is an operational strategy created to improve service and decrease costs of public transit (Smith et al., 2005). It is a control strategy that facilitates the movement of in-service transit vehicles through signalized intersections. In the simplest type of TSP, called passive TSP, the priority operates continuously, on the basis of the knowledge of transit routes and ridership patterns. Passive TSP does not require a transit detection or priority request, and it does not need any special hardware or software installations. It can be very efficient when transit operations are predictable.

The priority treatment can also be provided for transit vehicles following detection and subsequent priority request activation. This type of TSP is called active priority and it can be deployed in different manners within the specific traffic control environment. Active TSP can be achieved as unconditional or conditional. Unconditional active TSP provides priority treatment for every transit vehicle that sends a TSP request. Conditional TSP provides priority only to transit vehicles that meet certain conditions, such as running behind the schedule, or having a certain number of passengers on board. Active TSP can be implemented through the green extension, where the green time for the TSP movement is extended when a TSP equipped vehicle is approaching. This strategy only applies when the signal is green for the approaching transit vehicle. Another common strategy is the early green or red truncation strategy, which shortens the green time of the preceding phases to expedite the return to green for the transit phase. This strategy only applies when the signal is red for the approaching transit vehicle. Some other active TSP strategies in use are phase rotation, phase insertion, actuated transit phase, or a combination of strategies.

The most comprehensive TSP strategy is adaptive TSP. It takes into consideration the trade-offs between transit and traffic delay and allows adequate adjustments of signal timing by adapting the movement of the transit vehicle and the prevailing traffic conditions. It can also consider some other transit inputs, such as whether the transit vehicle is running on time or is late, the headway between two successive transit vehicles, and the number of passengers on board.

Another way to improve transit progression is to use some of the Predictive Priority Strategies (PPS) (Head, 1999; Langdon, 2002; **Zlatkovic et al., 2011). In general, PPS combines different TSP strategies and the communication between intersection controllers to provide a high-level of priority for transit vehicles with minimum disruptions for other traffic. This form of signal control for transit priority was first developed for trains on urban transportation networks. PPS uses a series of advanced detectors to track the vehicle that needs to be prioritized and allows a green signal progression for that vehicle at intersections. The signal controller in this case functions in accordance with the set of control logic commands that are activated once the transit vehicle is detected approaching the intersection. This is an adaptive traffic signal control strategy that allows the adjustment of signal phases to transit vehicles present in the intersection area. PPS application to rapid transit modes (Light Rail or Bus Rapid Transit) could achieve uninterrupted progression of these vehicles through the intersections, without waiting for the signal changes. So far, PPS has only been used for rail-transit modes.

The ASC/3 controller software has built-in TSP features for green extension and early green strategies. Custom defined TSP strategies can be achieved through the application of the ASC/3 logic processor. Control logics can be adjusted for different types of priorities for public transit.

This chapter presents the application of ASC/3 SIL in VISSIM simulation for an evaluation of TSP and PPS for Bus Rapid Transit (BRT), compared to the transportation network without any type of priority for transit vehicles. The goal of the chapter is to explore the capabilities of ASC/3 SIL software in providing different transit priority strategies. This is achieved through back-to-back comparisons and analysis of three different microsimulation model scenarios of a base case network, which is a planned BRT line in West Valley City, Utah. The chapter is organized in six sections. The following section describes the ASC/3 controller and its SIL applications in more details. The third section describes the network and simulation models. It is followed by the results and discussion sections. Finally, the major conclusions of the study are given in the last section.

11.2 ASC/3 Controller and Software-In-the-Loop Applications

The ASC/3 controller is the latest series of Advanced System Controllers manufactured by Econolite Control Products (Econolite Control Products, Inc., 2008). It offers a vast array of control, coordination, preemption and TSP features, extent detector options, and communication abilities. It is also able to support very complex signal timing settings through the Logic Processor. A total of 100 logic commands can be accessed directly in the controller, and an additional 100 logic commands can be enabled through a special extension file. These commands can control and combine all the controller features and emulate external logic that is not included in the default settings.

The ASC/3 controller has been frequently used for HIL simulations. However, since HIL simulation is very time and resource demanding, a better solution is found in the ASC/3 SIL application developed for VISSIM simulation software (Urbanik et al., 2006; Stevanovic et al., 2009). ASC/3 SIL runs from the same code base as the hardware controllers, and they perform identically. This application provides many opportunities for evaluating and analyzing traffic control strategies that could be performed within a simulation environment. Once all the tests have been done in the simulation, the control strategies can be easily transferred to the field controllers by simply uploading the database file created during the simulation. This saves time, effort, and costs that could be incurred if the changes and testing are performed on a field controller. Another big advantage of the ASC/3 SIL is that it can run ten times faster than the real time during simulation, which greatly reduces the time needed to test a scenario in VISSIM. The ASC/3 SIL is comprised of the Data Manager (or Database Editor), Traffic Control Kernel, Controller Front Panel Simulator, and VISSIM DLL Interface components (Zlatkovic et al., 2010). The Data Manager is an application for managing the controller timing data of the simulated controllers while in the Operating System (OS) environment. This software is more intuitive and easier to use than the controllers' normal front panel data entry screens. The database file for the ASC/3 SIL and an actual ASC/3 controller are identical. The Traffic Control Kernel is the virtual ASC/3 core software that operates under OS. It encompasses all internal processing that occurs between the mapped field inputs that are passed from VISSIM and subsequent calculation of commanded field outputs that are passed to VISSIM. This interface guarantees consistency in traffic control operation between the simulated ASC/3 SIL running under VISSIM and a physical ASC/3 controller. The Controller Front Panel Simulator is a Graphical User Interface (GUI) designed to simulate the 16 line × 40 character display and keypad found on the ASC/3 physical controller. This GUI permits the display of status and data along with the changing of all user data settings within the simulated ASC/3 controllers running under VISSIM. Any changes made to the controller settings are stored in the simulated controller's database. The VISSIM DLL interface couples the ASC/3 simulated controllers to VISSIM. It allows VISSIM to pass detector and other Input/Output functions to the simulated ASC/3 controllers and to receive back controller status information.

The ASC/3 controller offers built-in preemption and TSP functions. The latest version of the ASC/3 SIL has these options too, making it possible to test different priority strategies in simulation. Studies that looked into the ASC/3 SIL priority showed the capabilities of the software (Zlatkovic et al., 2010; Winters and Abbas, 2011; Qing et al., 2011).

Another ASC/3 option that has just begun to emerge in the SIL application is the logic processor. Logic commands offer additional external control logic that does not exist in the default settings. A study that used an ASC/3 SIL logic processor to evaluate phase termination based on traffic flow data under recurring congestion showed advantages of external control logic and the ability of ASC/3 SIL to apply user defined logic controls (Smaglik et al., 2011).

This study explores the capabilities of the built-in TSP strategies, but also in a greater manner the use of the logic processor for custom-defined PPS. A series of logic commands was developed in order to define extensive priority strategies, beyond those that are offered within the controller software.

11.3 Project Description

The methodology researched in this study was applied and tested on a real-world network with a planned BRT line through VISSIM microsimulation. This section describes the network and the scenarios that were used, with detailed explanation of how the PPS strategies were achieved through the ASC/3 logic processor.

11.3.1 Project Network

The network selected for this study is a part of a future BRT line along 5600 W Street in West Valley City, Utah. The planned 5600 W BRT line involves five miles of dedicated center-running BRT lanes from 2700 S to 6200 S, with a total of six BRT stations, as shown in Figure 11.3. This type of layout with center-running transit lines is convenient for analyzing different aspects of TSP, from both operational and safety points of view. The network was originally developed as a part of a research project to analyze traffic and transit impacts for the target year 2030, when big changes in land development, traffic, and transit patterns are expected (Martin and Zlatkovic, 2010). VISSIM models were developed, calibrated and validated for current traffic conditions in 2009, and projected traffic volumes for 2030. Here we apply the 2030 estimates with some small changes in traffic and transit patterns to make them more suitable for the focus of the research. Three modeling scenarios were used for the study: No TSP scenario, TSP scenario and PPS scenario. All scenarios are customized to work in the ASC/3 SIL simulation environment.

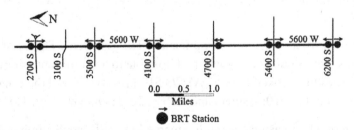

Figure 11.3: 5600W Base case network

11.3.2 No TSP Scenario

This scenario introduces the center-running BRT line without any special control treatment. The seven traffic signals are optimized in SYNCHRO for the 2030 volumes and road design, and these signal timings are incorporated in ASC/3 SIL. The headway for the BRT buses is set to 8 min in each direction. This is less than the planned 10-min headways, but it was changed in order to assess priority strategies in more detail. The duration of the VISSIM simulation was 2 h, for the 4:00 to 6:00 PM peak period, with a 15-min build-up time. The outputs from the simulation were averaged from ten simulation runs with different random seeds. All the same settings were used for the other two scenarios, with an addition of TSP or PPS.

11.3.3 TSP Scenario

This is an extension of the No TSP scenario. Green extension and early green (red truncation) strategies were defined using the built-in ASC/3 TSP features. For each intersection, the TSP settings were as follows:

- Maximum green extension for BRT phases: 10 s
- Maximum red truncation for conflicting through movements: 10 s
- Maximum red truncation for (all) conflicting left turns: 5 s

With these TSP strategies, the total gain for BRT buses was up to 20 s, depending on the moment during a cycle when the bus approached the intersection. TSP was defined as unconditional priority, which means that every BRT bus that was in the network sent a TSP request and was serviced accordingly.

11.3.4 PPS Scenario

This scenario is using custom-developed priority strategies achieved through a series of logic commands defined within the ASC/3 SIL logic processor. Four basic strategies were defined and simulated:

- Intersection communication
- Green extension
- Early phase termination
- Phase rotation

The main postulate of PPS was that none of the phases (vehicular or pedestrian) could be omitted, no matter which strategy was active at the time. This would provide normal intersection functions, with some modifications in operations when the priority was active.

Intersection communication is one of the postulates of the Predictive Priority. It means that the information about the presence of transit vehicles is sent from one intersection to the adjacent ones, giving them enough time to prepare for the approaching transit vehicle and serve it with minimum delay, and minimum impacts on vehicular traffic. The intersection communication could not be achieved directly with the ASC/3 SIL controllers. Instead, detectors for the downstream intersection were set at the previous one, simulating a signal that would be sent between the intersections. This signal was then delayed based on the spacing between the intersections, which is half a mile to one mile for the given network, and the presence of transit stops in the mid-block section. This signal would become active when the transit vehicle was 200–300 ft. from the intersection. It would then activate one (or a combination) of priority strategies, depending on the moment within a cycle when the vehicle appeared and the current phase timings at the intersection.

The green extension provides extra green time for a transit vehicle that is approaching an intersection, and it is estimated that it will not clear the intersection before the green ends. The build-in TSP strategies for the green extension work the same way, but in this case this was achieved through control logic. This logic works as follows:

IF
1. BRT detected AND
2. BRT phases timing green

THEN
3. Turn off minimum recall for all phases
4. Turn off detectors for conflicting phases
5. Call MAX 2 maximum green time for BRT phases
6. Set coordination free
7. Set green for BRT phases

The IF condition for this strategy is that a BRT bus is detected approaching the intersection, and the green time for BRT phases is currently on. The logic makes sure that the bus will clear the intersection before the green ends. The first step is to turn off detector actuations for all conflicting phases, and to turn off minimum phase recalls (if any). This will clear calls for conflicting phases and give an opportunity to the BRT phases to continue timing green. However, the duration of this green time can be constrained by the maximum phase green time, or the coordination offset. The ASC/3 controller has an option of defining three maximum green times, where MAX 1 is the standard maximum green, while MAX 2 and MAX 3 are optional, and they can be activated through the control logic. For the purpose of green extension, the logic refers to the MAX 2 time for the BRT phases, which is in this case defined large enough to allow the BRT bus to clear the intersection on green. To maintain the coordination offset, the controller can also end the green time of the coordinated phases at a certain point during the cycle. For the analyzed network, the coordinated phases at each intersection are the same as the BRT phases. This can conflict with the green extension, so the logic sets coordination to "free running" until the bus has cleared the intersection. Setting the control logic to dwell in green ensures that the BRT phases will remain green while the conditions are satisfied. When the bus crosses the stop bar, this logic will become inactive and the intersection will return to normal operations.

Early phase termination is the same strategy as early green or red truncation. If a BRT bus is detected approaching an intersection, and some of the conflicting phases are timing green at that moment, this strategy will terminate those phases to provide an earlier start for the BRT green phases. The logic that drives this strategy is as follows:

IF
1. BRT detected AND
2. Conflicting phase is timing green

THEN
3. Turn off detectors for that conflicting phase

When the detectors for the conflicting phase are turned off, the call for that phase will end and it will stop timing green once it reaches the minimum phase green time. It should also be noted that this strategy will not omit any phase, whether or not that phase is on a minimum recall. The logic becomes active once the phase green starts timing, which ensures minimum green for that phase. If one of the conflicting pedestrian phases, which time concurrently with the through movements, is active at the same time as this logic, the conflicting phases will end when the pedestrian phase turns red. It means that active pedestrian phases will not terminate earlier. Turning the conflicting phases' detectors off is a better option than forcing their green time to end (which can also be achieved through the control logic), because in this case the conflicting phases will gap out, which will not disturb intersection coordination, and is more fair to the vehicles on the conflicting movements.

Phase rotation is a strategy that changes the phase sequence in order to serve a transit phase faster. In this case, only the phases on the same intersection approach (within the same control barrier) can be rotated. Along the studied BRT corridor, the phase sequence is defined as leading left turns and lagging through movements for all intersections. All BRT phases time concurrently with vehicular through phases. If a BRT vehicle is detected at the intersection while the side street through movements have green, phase rotation will change the sequence for left and through phases at the main approach, allowing the through movements to be served first, and left turns after that. This strategy reduces delays for transit vehicles, but it can also have safety benefits in a case of a transit lane that is positioned in the middle of the roadway (especially for exclusive BRT or Light Rail Transit – LRT lanes). It reduces conflicts between transit and left turning vehicles. The logic behind this strategy is as follows:

IF
1. BRT detected AND
2. Left turns on BRT approach timing red

THEN
3. Select alternative sequence with leading through and lagging left phases

If a BRT bus is detected, the second IF command checks the timing for the left turn phases on the main (BRT) approach. If these left turn phases are red at the moment, two options are possible: either the through phases on the main approach (and BRT phases) are green, or any phases (left or through) on the side approach are green. In the first case, if the BRT phases and the concurrent through movements are green, the bus will clear the intersection and deactivate phase rotation. However, if some of the side street phases are green at the moment, it means that both left turns and through (and BRT) movements on the main intersection approach are timing red. The normal phase sequence on the main approach in this case would start with leading left turns on the main approach, and will lag through and BRT phases. But in this case the logic will be active and it will select an alternative sequence, which is defined as leading through phases and lagging left turns, serving the BRT phases first. This alternative sequence has to be pre-defined in the ASC/3 SIL configuration and referred to through a proper logic command. The early phase termination strategy will always be active along with phase rotation.

Depending on the moment when a BRT vehicle is detected approaching an intersection and current phase timings, either one or a combination of strategies will become active, giving a certain priority to the BRT vehicle. As in the previous scenario, priority for the BRT vehicles was unconditional. Figure 11.4 shows an example of applying ASC/3 logic processor in PPS programming.

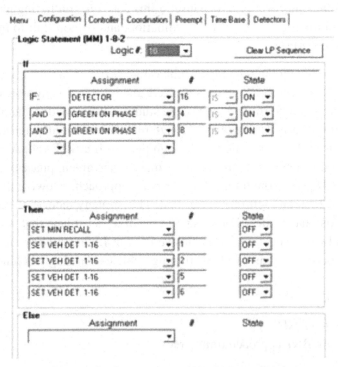

Figure 11.4: ASC/3 logic processor GUI: PPS application example

11.4 Results

For the purpose of evaluating different priority strategies, VISSIM was coded to record travel times (vehicular and BRT), intersection performance, signal phase timings, and overall network performance. The results were collected for each scenario and then compared.

11.4.1 Travel Times

Travel times for vehicles and BRT buses were measured for segments between each pair of signalized intersections, in the northbound and southbound direction. A comparison of the average travel times for the 2-h simulation period between scenarios is given in Table 11.1.

Table 11.1: Travel times for BRT and vehicles (in seconds).

SB	NO TSP		TSP		PPS	
Segment	BRT	Cars	BRT	Cars	BRT	Cars
2700 S – 3100 S	135	61	114	61	109	64
3100 S – 3500 S	69	71	66	71	59	81
3500 S – 4100 S	209	111	193	111	173	130
4100 S – 4700 S	199	113	173	113	175	132
4700 S – 5400 S	187	127	184	127	171	128
5400 S – 6200 S	196	119	181	119	169	128
Total	995	602	912	602	856	663
NB	NO TSP		TSP		PPS	
Segment	BRT	Cars	BRT	Cars	BRT	Cars
6200 S – 5400 S	141	135	141	136	145	139
5400 S – 4700 S	223	118	207	119	177	132
4700 S – 4100 S	136	136	125	134	111	131
4100 S – 3500 S	163	160	164	162	145	147
3500 S – 3100 S	87	64	87	64	91	71
3100 S – 2700 S	59	70	58	70	58	72
Total	809	683	782	685	727	692

Table 11.2: 4100 S intersection performance comparison.

Movement	NO TSP			TSP			PPS		
	Vehicle	Delay (s)	Stops	Vehicle	Delay (s)	Stops	Vehicle	Delay (s)	Stops
NBT	334	29.5	0.79	333	28.1	0.79	333	27.6	0.64
NBL	131	44.0	0.90	130	49.1	0.93	130	64.6	1.00
SBT	926	12.2	0.24	923	10.9	0.22	918	24.9	0.50
SBL	186	80.0	1.06	187	77.7	1.06	189	70.4	1.00
EBT	824	41.5	0.80	823	43.1	0.82	832	46.3	0.86
EBL	174	40.9	1.14	174	42.8	1.16	176	43.2	1.18
WBT	895	41.8	0.81	894	43.2	0.82	905	46.4	0.85
WBL	169	37.5	1.09	168	39.8	1.11	170	41.0	1.15
BRT NB	8	46.7	0.81	8	37.0	0.63	8	21.8	0.11
BRT SB	7	64.7	0.99	8	47.7	0.85	8	28.5	0.36
Total	3653	34.9	0.71	3647	35.4	0.71	3669	40.7	0.79

SB, southbound; NB, northbound; WB, westbound; EB, eastbound;
L, left movement; T, through movement.

11.4.2 Intersection Performance

The intersection performance parameters, such as vehicle and person delays, stop delay, number of stops, and average and maximum queues were measured for each movement at each intersection. The example shown in Table 11.2 is for the intersection of 5600 W and 4100 S, which is the intersection in the middle of the network. The comparison is given for the number of vehicles, vehicle delays and number of stops per vehicle for the 5:00–6:00 PM peak hour. Table 11.3 shows weighted performance measures on the intersection level for all

Table 11.3: Network level intersection performance.

Mode	Intersection	Number of Vehicles			Delay Per Vehicle (s)			Stops Per Vehicle		
		No TSP	TSP	PPS	No TSP	TSP	PPS	No TSP	TSP	PPS
Cars	2700 S	7805	7805	7802	28.1	28.2	29.6	0.82	0.82	0.82
	3100 S	7104	7105	7100	31.6	32.1	32.9	0.67	0.68	0.70
	3500 S	10173	10170	10121	35.9	36.7	43.9	0.86	0.86	0.92
	4100 S	8713	8708	8740	31.9	32.6	38.5	0.73	0.73	0.83
	4700 S	7237	7236	7233	29.7	30.0	33.4	0.74	0.76	0.88
	5400 S	8313	8316	8305	32.4	33.2	36.1	0.81	0.82	0.83
	6200 S	7896	7902	7919	30.2	30.4	33.3	0.75	0.75	0.77
BRT	2700 S	30	30	30	29.0	25.3	21.8	0.32	0.27	0.13
	3100 S	30	30	30	17.5	7.0	6.5	0.31	0.17	0.15
	3500 S	30	30	30	35.4	35.0	21.9	0.71	0.62	0.14
	4100 S	30	31	31	54.6	41.5	25.5	0.88	0.72	0.25
	4700 S	30	30	30	60.7	40.4	25.6	0.61	0.35	0.34
	5400 S	30	30	30	31.7	30.9	25.9	0.32	0.38	0.26
	6200 S	29	29	29	49.4	35.3	24.9	0.62	0.47	0.26

Table 11.4: Signal-phase durations (in seconds)

Signal group	No TSP			TSP			PPS		
	Green	Yellow	Red	Green	Yellow	Red	Green	Yellow	Red
1 – SBL	14.5	3.0	112.5	14.2	2.9	112.9	15.9	3.2	110.8
2 – NBT	47.3	4.5	78.1	48.9	4.5	76.5	47.1	4.9	78.0
3 – WBL	8.5	2.9	118.6	8.5	3.0	118.5	8.4	2.9	118.7
4 – EBT	38.3	4.0	87.7	36.9	4.0	89.1	36.4	4.0	89.6
5 – NBL	11.2	2.9	115.9	10.7	2.8	116.5	9.9	2.9	117.2
6 – SBT	50.8	4.5	74.7	52.6	4.5	72.8	54.1	4.6	71.3
7 – EBL	8.1	2.9	118.9	7.8	2.8	119.4	8.8	2.9	118.3
8 – WBT	38.6	4.0	87.4	37.9	4.0	88.1	36.0	4.0	90.0

intersections in the network and the entire analysis period (4:00–6:00 PM). The results are given separately for private cars and BRT vehicles.

11.4.3 Signal Phase Timings

TSP strategies can impact phase timings, especially green time durations. In order to assess these impacts for each of the three examined strategies, VISSIM was coded to provide signal status during each 0.1 s. Table 11.4 shows an example of average phase time durations during a cycle for each scenario for the intersection of 5600 W and 4100 S.

11.4.4 Network Performance

Impacts and benefits of the different priority strategies can be assessed on a network wide level. Table 11.5 presents a network performance comparison for the most relevant parameters.

11.5 Discussion

The assessment of results on different levels (corridor, intersection, network-wide) provides more insight on how TSP and PPS strategies impact traffic and transit operations. This section provides detailed analysis of results and comparisons among the tested scenarios.

Table 11.5: Network performance.

Parameter	NO TSP	TSP	PPS
Average delay per vehicle (s)	57.9	58.7	65.2
Average stopped delay per vehicle (s)	42.0	42.7	48.2
Average number of stops per vehicles	1.4	1.4	1.5
Average speed (mph)	23.7	23.6	22.7

11.5.1 Travel Times

The travel time results for each scenario show greater BRT travel times in the southbound than in the northbound direction. This is expected, because southbound is the PM peak direction with more transit riders and greater station dwell times. An implementation of different TSP strategies improves BRT travel times. The results show that the green extension/early green strategies reduce BRT travel times by 8% in the southbound and 3% in the northbound direction when compared to the No TSP scenario. PPS strategies result in even more travel-time savings for BRT vehicles: 14% in the southbound and 10% in the northbound direction.

Green extension/early green strategies have no impact on vehicular travel times along the main corridor. However, PPS strategies tend to increase vehicular travel times in the southbound direction by approximately 10%. These travel times are impacted by the phase rotation and disturbances in intersection coordination caused by some of the PPS strategies.

11.5.2 Intersection Performance

An analysis on the intersection level shows that the green extension/early green strategies have certain benefits, while PPS offers significant savings in delays and number of stops for BRT in both directions (see Tables. 11.2 and 11.3). Built-in TSP reduces BRT delays in the range between 1% (at 3500 S) and 60% (at 3100 S). Reductions in BRT delays in the PPS scenario vary from 18% (at 5400 S) to 63% (at 3100 S).

TSP strategies have minimal impacts on vehicular traffic along the main corridor and on side streets. Along the main corridor, PPS causes an increase in delays mostly for vehicles on southbound through movements and some left turns. These movements are affected by the phase rotation and impact that PPS has on coordination. Some smaller impacts of PPS are noticed on side street movements. The increase in car delays caused by PPS varies from 4% (at 3100 S) to 22% (at 3500 S).

11.5.3 Signal Phase Timings

TSP and PPS have no major impacts on green phases' durations, as given in Table 11.4. However, the distribution of green times changes slightly with different strategies. Both TSP and PPS increase green times for through movements along the corridor. Green times are generally decreased for all other movements along the corridor and on side streets. It can also be observed that the green phase durations for some left turns are impacted by the phase rotation strategy in PPS.

11.5.4 Network Performance

Network performance results given in Table 11.5 are similar to the single intersection results. It can be seen that TSP has no major impacts on the network-wide level performance, whereas PPS increases average delays per vehicle by about 12%. The reason for this is the same as in the case of a single intersection (phase rotation and impacts on coordination).

All the compared parameters show same impacts/benefits that different strategies have on vehicular traffic and BRT.

11.6 Conclusions

The main goal of this chapter is to explore the capabilities of the ASC/3 SIL software in analyzing different types of transit priority strategies, through the built-in ASC/3 TSP features and custom-developed priority achieved through the logic processor. This chapter shows that the ASC/3 SIL has proven to be a very powerful tool for this type of analysis. It means that for a real network, these analysis can be performed in a simulation environment, removing the risk of any errors that could be made in an on-site controller programming. The ASC/3 SIL has an option of creating a database file that can be directly transferred into a field controller.

The results from the base-case network are hypothetical, because they are given for assumed transit operations. In order to record different aspects of the defined priority strategies, transit frequencies were increased beyond those that would be implemented in the planned network. This increased the impact that transit and TSP had on vehicular traffic. However, the results are significant because they can offer some guidelines for defining optimal priority strategies. In this chapter, they are used to show the extent of the ASC/3 TSP features and user-programmable priority strategies. It is demonstrated that SIL can be applied to real-life transportation networks and used for traffic optimization purposes.

The main contribution of this work is that it provides a set of instructions for different levels of TSP that can be directly programmed into the field traffic controllers, without the need to install new hardware or software. The analysis was performed for ASC/3 controllers, but it can be easily customized for any other type that supports TSP options and/or logic processor.

Some of the topics for future research in this area can be the following:

- A combination of built-in TSP features and logic processor to optimize transit priority strategies for a given transportation network
- Application of the logic processor to conditional and adaptive transit priority
- Application of the logic processor in resolving two or more conflicting priority requests
- Application of the logic processor in analyzing traffic control strategies that are beyond standard operations

References

Econolite Control Products, Inc., 2008. Advanced System Controllers ASC/3 Programming Manual.

Head, L., 1999. Improved Traffic Signal Priority for Transit. Transit Cooperative Research Program Interim Report No A-16A, Transportation Research Board of the National Academies, Washington, D.C.

Langdon, S.M., 2002. Simulation of Houston light rail transit predictive priority operation. ITE J. 72 (11), 28–32.

Martin, P.T., Zlatkovic, M., 2010. Evaluation of transit signal priority strategies for bus rapid transit on 5600 West Street in Salt Lake County, Utah. Technical Report MPC-09-213A, Mountain Plains Consortium.

Qing, H., Head, L., Ding, J., 2011. A heuristic algorithm for priority traffic signal control. Transportation Research Board 90th Annual Meeting, Washington D.C., USA.

Smaglik, E.J., Savolainen, P.T., Steele, R.C., DiBiasi, J.E., et al. 2011. Delay Analysis of a Traffic Signal Phase Termination Algorithm Using Computer Simulation. Transportation Research Board 90th Annual Meeting, Washington D.C., USA.

Smith, H.R., Hemily, B., Ivanovic, M., 2005. Transit Signal Priority (TSP): A Planning and Implementation Handbook. Intelligent Transportation Society of America, Washington, D.C.

Stevanovic, A., Abdel-Rahim, A., Zlatkovic, M., Amin, E., 2009. Microscopic Modeling of Traffic Signal Operations: Comparative Evaluation of Hardware-in-the-Loop and Software-in-the-Loop Simulations. Transport. Res. Rec.: J. TRB 2128, 143–151, doi:10.3141/2128-15.

Urbanik II, T., Venglar, S. P., 1995. Advanced Technology Application: The "SMART" Diamond. Institute of Transportation Engineers 65th Annual Meeting. Denver, Colorado, USA, August 5 – 8, 1995. Compendium of Technical Paper, p. 164-168.

Urbanik II, T., Kyte, M., Bullock, D., 2006. Software-in-the-loop simulation of traffic signal systems. SimSub Mid-year Newsletter, 15–19.

Winters, K.E., Abbas, M.M., 2011. Optimization of transit signal priority extensions along an uncoordinated arterial. Transportation Research Board 90th Annual Meeting, Washington D.C., USA.

Zlatkovic, M., Martin, P.T., Stevanovic, A., 2010. Evaluation of transit signal priority in RBC and ASC/3 software-in-the-loop simulation environment. Transportation Research Board 89th Annual Meeting, Washington D.C., USA, 1.

Zlatkovic, M., Martin, P.T., Stevanovic, A., 2011. Predictive Priority for Light Rail Transit: University Light Rail Line in Salt Lake County UT. Transport. Res. Rec. J. TRB 2259, 168–178, doi:10.3141/2259-16.

CHAPTER 12

Mobility Network Evaluation in the User Perspective: Real-Time Sensing of Traffic Information in Twitter Messages

Zafeiris Kokkinogenis*, João Filguieras**, Sara Carvalho†, Luís Sarmento‡, Rosaldo J.F. Rossetti*

*LIACC, DEI, Faculdade de Engenharia da Universidade do Porto, Porto, Portugal; **Instituto de Engenharia de Sistemas e Computadores, INESC-ID, Lisbon, Portugal; †MIEIC, SAPO Labs, Faculdade de Engenharia da Universidade do Porto, Porto, Portugal; ‡LIACC, SAPO Labs, Faculdade de Engenharia da Universidade do Porto, Porto, Portugal

12.1 Introduction

Modern cities have become centers of employment and trade whereas, on the other hand, losing the main bulk of their populations to peripheral areas. This has caused a massive increase in traffic volumes on roads and transportation services as people are forced to commute daily between their homes and their workplaces. For commuters this means a considerable amount of time wasted every day and therefore a decrease in quality of life. It becomes important and necessary to optimize our existing mobility systems. The study of an urban network involves the placement of several types of sensors and devices along the network to gather data. Such data is used to calculate measures, such as the *traffic density* and *flow*, which allow for a better understanding of the network at a given moment. This information is then used for improving the traffic flow and organization of the urban spaces. The sensors and devices used to collect data include hardware such as cameras, inductive loops and radars. Even though they are effective in their respective functions, these sensors have some limitations. First, they are relatively expensive and require constant maintenance. Second, they are not mobile and cover only a restricted area of the network. Third, these sensors tend to be very specialized, that is, they only collect data of one particular type (e.g., vehicle count). Finally, the information they collect is usually property of one organization or company and it is usually difficult to obtain access to it, even for the purpose of research.

However, the aforementioned traditional approaches offer us approximate and quantitative metrics such as traffic volumes and wait times, but fail to provide a subjective user

Advances in Artificial Transportation Systems and Simulation.
Copyright © 2015 Zhejiang University Press Co., Ltd. Published by Elsevier Inc. All rights reserved.

perspective of the performance of a transportation network that we consider important to understand the inherent dynamics. We believe that in order to provide solutions to tackle mobility issues and to devise new urban designs, we need to complement objective data with subjective data. We share the idea that subjective data, such as opinions, contributes to the overall knowledge on a subject (Tsytsarau and Palpanas, 2012). Carvalho et al. (2010) was one of the first works to propose exploring an alternative mobility information source that can be useful to the traffic characterization problem, and simultaneously, overcomes some of the limitations of the traditional sensors described earlier: Twitter messages, called tweets.

Twitter is one of the applications that has experienced an increasing popularity among Internet users and has deeply transformed the way people communicate around the world. The Twitter community has been growing at a considerable pace since its creation and its role has become very important in many social, economic, and even political contexts. For example, Twitter was used to report the conflicts that followed the Iranian Presidential election in 2009, despite the active censorship performed by the Iranian authorities over more traditional media. Twitter was also used by the victims of the fires in California, in 2007, and Victoria, Australia, in 2009, to report accurate information in real-time and to help other victims.

Likewise, due to their real-time and ubiquitous nature, it is believed that Twitter messages can be used for a variety of other purposes, for example, as a complementary source of relevant and up-to-date traffic information. In fact, anonymous users sometimes post messages that describe relevant traffic information, such as: *This traffic jam doesn't seem to have an ending. I'm still stuck in EN12.*

Therefore, the identification of this type of messages may be very useful for traffic characterization purposes, especially because Twitter allows overcoming some of the problems that exist with other sensors. First, there are virtually no costs involved: these are software sensors that require practically no maintenance, since users voluntarily communicate information. Second, Twitter-based sensors allow obtaining information from potentially every point of the network, even those that are far from the main traffic axis. Third, Twitter users can describe a wide range of traffic related events, which go beyond simple frequency counts. Finally, the information gathered is open to the public and can be accessed freely.

Official traffic information sources also use Twitter to broadcast information. Several Twitter "users" are in fact robots from news agencies the inject messages containing traffic information in the Twittosphere, such as: *08:53, 29-04-2010, IP4, km 97*, direction west, accident, drive safely. Obviously, these robot-sent messages are not interesting from the point of view of sensing traffic information (they report information previously sensed and analyzed). The problem we address in this chapter is the identification of messages relevant to the traffic characterization sent by anonymous users (i.e., not by robots). He et al. (2013) analyze the way in which social media can be used to improve traffic conditions forecasting. The conclusion of their work is a strong correlation between traffic measurements and traffic tweets leading to an improved performance in traffic prediction. Indeed, the motivation behind this work is the observation that people tend to share online their traveling experience posting traffic-related content in social media.

Following this trend, our goal is to discuss a methodology to support decision-making in transportation domain, built upon the tweet-traffic classifier proposed by Carvalho et al. (2010) as integrated part of the methodology. The intention is to go beyond the prediction purpose and create a tool for participatory sensing. Following this idea, we discuss a scheme for an information system capable of retrieving public opinion on urban and transportation mobility through social media. The goal of this system is to provide strong indicators of user sentiment toward the efficiency and quality of transport networks, not only as a whole, but also for specific routes and specific attributes. These indicators give us the perceived quality of the mobility network by its users, and enable us to evaluate how our idea of the users' needs actually matches what they desire. Urban planners can then make decisions on the basis of subjective opinions as well as objective metrics. Furthermore, it becomes possible to study the impact of network changes to users on more direct ways than simply watching for variations in traffic volumes or delay times. From a mobility researcher standpoint, sentiment also helps to characterize routes, as we try to understand why they are popular. Sentiment comes as another contribution to the semantic enrichment of routes. Compared to traditional approaches of gathering traffic data that require dedicated hardware such as sensors, this system is inexpensive, making it a viable addition to existing platforms at a very low cost.

In this chapter, we will focus on the second stage of the discussed methodology, the classification step. There are the two main challenges in this stage. First, from the text classification point of view, this problem is highly unbalanced. By manually sampling a large collection of Twitter messages, we were able to estimate that the percentage of traffic messages sent by anonymous users (i.e., excluding messages automatically posted by official traffic agencies) correspond to less than 0.05% of the messages at stake. This means, that it should be extremely difficult to achieve high precision in the identification of relevant Twitter messages. Secondly, as a consequence of this highly unbalanced distribution, we face the additional challenge of creating an appropriate and representative dataset for training the classifiers. With such low ratio of positive cases, manual annotation of a balanced corpus becomes unfeasible, so alternative strategies have to be devised. In this chapter, we focus on these two challenges.

The following parts of this chapter are structured like this: Section 12.2 reviews the concept of using social media, particularly the twittosphere, as mean to sense people daily activities and interaction. We present how classification and sentiment analysis of tweet messages has become an appealing approach in traffic detection and prediction, but also in supporting planners in their policy-making processes. Next, we show a picture of a methodology that uses twitter to support decision-making in transportation domain as feedback to users experience and opinion. In Section 12.5 we discuss the experiment set-up and the results of the classification process as integrated part of the proposed methodology. Finally we give our conclusion about leveraging existing traffic information using traffic-related tweets and the users' perspective of several aspects of the transportation network.

12.2 Social Media as Artificial Sensors

The use of social media has been growing. Twitter, in particular, due to its simplicity, is a popular platform where users tend to publish their emotional states, ideas and other experiences, almost in real-time. This creates vast amounts of user-generated content, that can cover many topics, and that have attracted the attention of many researchers.

Sakaki et al. (2010) present a system that uses Twitter as a social sensor for the real-time sensing of messages that report earthquake related events. This system is able to detect, with high probability, most of the earthquakes, with a seismic intensity scale greater than two, reported by the Japan Meteorological Agency, just by monitoring Twitter messages.

Asur and Huberman (2010) demonstrate how social media, in particular Twitter, can be use to predict real-world outcomes. The study focuses on the prediction of box-office revenues for out-coming movies. In the end, they conclude that a simple model that senses tweets on a particular topic can outperform some market-based predictors, therefore proving the forecasting power of social media. A similar study has been carried on by Tumasjan et al. (2010), focusing in the predictions of Elections. They conclude that the mere number of tweets reflect voter preferences and comes close to election polls, and that the tweets are not only about spreading political opinions, but also to discuss these opinions with other users.

The finding of high-quality content in social media is studied by Agichtein et al. (2008). In the chapter, they investigate methods that allow the identification of high-quality content automatically by exploiting community feedback, such as links between items and explicit quality ratings from members of the community.

12.2.1 Twitter in Traffic and Transportation domain

As declared in the previous section, nowadays, a trend has been created that tries to explore and exploit the rich information in social media in various application domains. Intelligent Transportation Systems is one of them. In the last 3 years, significant works have been published studying ways to leverage traffic information with user-generated content to provide solutions for tackling mobility issues and to devise new services.

Indeed, Sakaki et al. (2012) developed a system to provide valuable information for drivers extracted from social media. The approach authors followed is based on the collection of tweets referring to target events by directly applying tweet SVM classifiers. In the second stage, the system extracts location information from each tweet using GPS information, geo-location web services, a user-generated dictionary, and contextual information. Authors concluded the method can classify tweets referring to heavy traffic from Twitter with about 0.87 precision and can extract location information from those tweets with 0.85 precision.

Sasaki et al. (2012) performed a feasibility study on detecting train operational status exploiting Twitter as a sensor. Authors manage to obtain train status information with a high degree of both the accuracy and the real-time detection.

Albuquerque et al. (2012, 2013) discuss basic requirements for *proactive* real-time monitoring applications. Authors propose a framework to monitor moving objects and explore trajectory semantics that are sensitive to edynamics. Their works outline some of the features of a prototype application to monitor a fleet of delivery trucks, using Twitter message streams, from traffic authorities, to detect changes in road conditions, and to geo-reference the relevant facts.

12.2.2 Opinion Mining and Sentiment Analysis in Policy-Making

Opinion Mining is a growing and promising research field, especially using social media. As of late, researchers have taken a special interest in Twitter because of the diversity of its topics and the sheer amount of opinionated messages. Liu offers a comprehensive introduction to the fields of Sentiment Analysis and Opinion Mining (Liu, 2012). His work exposes and discusses the most widely studied sentiment topics and classification methods. Tsytsarau and Palpanas (2012) present a thorough review of the most popular algorithms for sentiment extraction in the literature and discuss their precision. Schweitzer (2012), a study on the unsolicited comments about airline services and transit agencies on Twitter is presented. Carpenter and Way (2012) propose an algorithm that performs sentiment analysis over time using data from Twitter, and describe a web-based application that generates sentiment-tracking graphs. Esparza et al. (2012) investigate the way user-generated micro-blogging messages can be used as a new source for recommendation systems based on extracted opinions. Chesley et al. (2006) focus on the automatic identification and extraction of opinions, emotions, and sentiments from text and multimedia.

In the recent years, we have been witnessing the explosion of what is usually called participatory sensing. Ordinary people take a proactive role in publishing comments and complaining online, increasingly using technology to record information about events and problems in all dimensions of their political and social life. Data collection and opinion mining approaches are seen as the cornerstones of large-scale collaborative policy-making. Stylios et al. (2010), present a method for extracting citizen opinions about governmental decisions from social media, as well as a technique for classifying opinion phrases in terms of their sentiment orientation. Additionally, authors define a metric for quantifying the impact of citizen opinions on governmental decisions, so that the former can be successfully used in subsequent governmental regulations. Kaschesky et al. (2011), a discussion about the advantages of opinion mining over surveys is offered, when dealing with citizenship issues. The authors propose an opinion mining approach that collects and analyzes citizen arguments and concerns. Lin et al. (2006) argue the possibility of providing support for decision makers to automatically track attitudes and moods in online media and user-generated content. Opinion mining and sentiment analysis have been proposed as key technologies in eRulemaking, allowing the automatic analysis of the opinions that people submit about pending policy or regulation proposals (Kwon et al., 2006; Shulman et al., 2005).

From the above literature review we conclude that tweets can be sensors of social dynamics that take place in people daily activities. The real-time testimony of on-going phenomena and their opinions and expectations can be used to identify problems with their environment and design solutions. This idea that has been applied to policy-making can be transposed to transportation domain in general.

12.3 Methodology

Next a general methodology is discussed to support decision-making in transportation domain as feedback to users experience and opinion. The idea behind this proposal is that indications of user sentiment toward the perceived efficiency and quality of transport networks enable transportation planners and managers to evaluate how their idea of the users' needs actually matches what they desire. To gather and analyze mobility-related opinions from social media we propose a pipeline with the following steps: (1) fetch user-generated documents, such as tweets, from social media, (2) determine which are related to mobility, (3) extract mentions of locations, and (4) analyze sentiment.

12.4 Fetching Data

Fetching public user-generated content does not require in most cases complex solutions and can be accomplished using a crawler or a specific API provided by the chosen web platform. Twitter, for instance, provides an API and restricts access for simple web crawlers. It also enforces a limit to the number of requests any given API client can make. Nevertheless, there are specialized crawlers, such as TwitterEcho (Boanjak et al., 2010), that are able to gather tweets in a fast and reliable fashion, while respecting the aforementioned restrictions.

A more complex problem is to find the small fraction of user-generated content of interest. While the mobility classifier presented in Section 12.4 already explicitly performs this task, it simply is not computationally sound to apply the classifier to every text, post, and message or tweet present in a social network. Some limits must be imposed when fetching content.

Since we are only interested in data related to mobility, a starting point would be any sort of conversation or explicit group of messages within that topic. On Twitter we can start by fetching tweets that contain hashtags or keywords we consider mobility related.

Certain users are robots, generally conveying information found elsewhere. Traffic-related robots also exist, relaying real-time information about traffic conditions on a given city, state, or country. Their feeds can be used as another starting point. Furthermore, followers of these feeds can also be of interest, especially because we have a strong indication that (1) they care about traffic or mobility and may express their opinions about it, and (2) they reside in a zone we are studying.

12.5 Sensing Mobility – Tweet Classification

Given a stream of messages from Twitter, that is, tweets, our goal is to identify the messages that contain relevant information for traffic characterization. This can, therefore, be seen as a binary text classification problem. There are multiple algorithms for performing text classification, which, given the appropriate training set, achieve relatively high performances. One of the main challenges in this setting is precisely that of generating a representative training set, since percentage relevant messages in the Twitter stream is extremely low ($< 0.05\%$). Thus, manually selecting positive examples among the messages available for training becomes an unfeasibly laborious task.

However, as explained before, there are several official agencies injecting valid traffic messages in the Twittosphere, using robot users. Robot users are quite simple to identify since their messages have very strict formatting. On the other hand, Twitter messages posted by human users tend to be very loose in terms of grammatical structure, and usually contain many spelling mistakes, nonstandard punctuation, and emoticons among other idiosyncrasies. Nevertheless, robot-generated messages provide good examples of vocabulary related to the description of traffic-related events (e.g., names of routes or critical locations in the network), which should be shared by many of the human-generated messages. Therefore, our strategy consists in using robot-sent messages – which are relatively easy to identify – for gathering an initial set of positive examples needed for training a classification model. Given the very small fraction of Twitters related to traffic, a balanced number of negative examples can be randomly chosen from the entire collection of Twitter messages.

After training the classifier with robot-sent messages (as positive examples) and messages randomly chosen from a collection of Twitter messages (as negative examples), we obtain a first version of the classifier. We then proceed by following a boot-strapping approach: we use the first version of the classifier (just described) to find additional positive and negative examples that can be used enrich the initial training set for training a better classifier.

Among the examples classified as relevant traffic messages by the first classifier we should have:

1. Messages generated by human users that are, in fact, relevant traffic messages, and that are interesting from the point of view of traffic sensing;
2. Message about traffic event, but which are sent by robot-users and, thus, not interesting for traffic sensing purposes;
3. Messages that are not related to traffic, that is, they have been incorrectly classified in terms of topic.

We add (1) as additional positive examples to the training set, while (3) are used as negative training examples. The messages in (2) are not included in the expanded training because they can neither be considered negative examples (they are traffic-related), nor they can be considered positive examples (they do not represent the messages we want to capture). Using such expanded training set, we train a second version of the classifier, which is expected to be more robust and to achieve higher precision in identifying relevant traffic messages posted by human users.

12.6 Named Entity Recognition

Once we have a sample of messages related to mobility, we can go even further and link each message to one or more locations. It is worthwhile to note that on Twitter a user may associate a location with each tweet, making this step easier. If this information is not made available the alternative is to use a Named Entity Recognition (NER) system to extract references to locations. A NER system processes a textual document and recognizes words or sentences that mention entities. Entities could be, for instance, locations, people, or organizations. A widely used system is the Stanford NER (Finkel et al., 2005). This system uses several word-level and sentence-level features to find references to entities and infer its type, for example, location, person, or organization.

However, a reference may be ambiguous. For example, two streets may have the same name in different cities, or even countries. As such, our next step is to disambiguate these references in a process called Entity Linking (EL). Our objective is to link one reference to one concept, in our case, a unique location, in a knowledge base. Table 12.1 shows an example using Wikipedia as the underlying knowledge base. Rao et al. (2013) propose a system that tackles

Table 12.1: Example of location extraction.

Step	Result
Raw tweet	I was stuck in traffic at the Brooklyn Bdg for hours #nytraffic
After NER	I was stuck in traffic at the Brooklyn Bdg <Location> for hours #nytraffic
Disambiguation	Brooklyn Bdg > Brooklyn_Bridge

this problem. The system starts by selecting a pool of candidate concepts using string similarity between the reference and the title of concept entries on the knowledge base. Afterward it performs a deeper comparison of each candidate with the reference and its context to find the correct match.

Named Entity Recognition and Entity Linking are more challenging on social media where messages are short and informal. These messages lack the amount of textual context that NER and EL systems rely on. Nevertheless, approaches for NER and EL on Twitter have already been proposed (Ritter et al., 2011; Meij et al., 2012). Ritter et al. shows that traditional NER systems perform poorly on Twitter and propose a system to overcome these difficulties by trusting less on things such as capitalization and punctuation. Meij et al. propose a system to perform Entity Linking on Twitter. They take into account that textual references to concepts are more likely to be abbreviated or misspelled on microblogs.

In our particular case we choose knowledge bases for specific mobility networks. For instance, if we wish to study any given city, we use a knowledge base containing entries for streets and other points of interest in that city. The richer the knowledge base the more accurate our disambiguation system will be.

In addition to references in the message, we can also consider the general location of a user, either via profile information, or employing geo-tagging techniques. Cheng et al. (2010) propose a system that estimates the location of authors based on their tweets. This system uses lists of words linked with specific locations that are automatically compiled. Since it uses only the contents of tweets themselves, it does not require any context information.

By the end of this step, we will have either a number of locations referenced by the user or his general location indicated by his profile or by geo-tagging techniques. In the case of Twitter, we can additionally have a fine-grained location specified by the user at the time of publication.

12.7 Sentiment Analysis

After linking texts to locations, we can use the sheer volume of mentions to perform an analysis of its importance. However, we can also apply sentiment analysis to have a better understanding of why these locations are mentioned. Sentiment-analysis systems process a textual document and estimate the polarity of any present opinions. This polarity may be estimated regarding the overall sentiment of a document, or, it may be estimated for an opinion toward a target referenced in the document. For instance, a document may reference a particular street X and it may express a negative opinion toward that street. In this case our sentiment analysis system will return a negative polarity toward the target, street X. Naturally sentiment polarity can be positive, negative, or neutral. While our main interest would be to detect polarity toward particular points of a mobility network, we can still consider polarity toward the whole

Table 12.2: Possible aspects to be considered.

Aspects	Keywords
Road conditions	"holes", "flood", "pavement"
Delay times	"slow", "stop", "jam"
Tolls or fares	"cheap", "costly"

network if only a generic location is available. Our analysis of expressed opinions can be taken one step further. Instead of polarity toward a target point of interest, we can also detect polarity toward aspects of target points of interest, such as road conditions or time delays (see Table 12.2).

Kobayashi et al. (2007) model opinions as tuples composed of an opinion holder, an opinion target, an attribute of the target being mentioned, and the polarity of the opinion. The authors then propose a machine learning system that uses contextual clues to infer polarity toward aspects of the target.

To implement a sentiment analysis system there are several proposed approaches for different types of documents. Following a recent trend there is a wide variety applied to Twitter as well. Thelwall et al. (2010) present a sentiment analysis system, available online[1]. This system employs a number of features, for example sentiment lexicons (i.e., dictionaries of positive and negative words), negation awareness to invert the polarity of negated words and emoticon lists. These features are used to train a machine learning classifier using an annotated dataset of MySpace comments. Bifet and Frank (2010) show a similar system for tweets and experiment with different machine-learning algorithms.

Selecting one of the approaches proposed in the literature for this particular domain is a matter of experimentation to find the approach that yields the best results. From a single tweet we may retrieve more than one opinion, because more than one aspect or location may be mentioned. An example of the end result of the pipeline is shown in Table 12.3.

Table 12.3: Processed examples.

Raw tweet	Location	Aspects
I was stuck in traffic at the Brooklyn Bridge for hours #nytraffic	*Brooklyn_Bridge*	Delay times (negative)
Almost lost a wheel on A2 because of a hole!	A2_road_(Great_Britain)	Road Conditions (negative)
I've got home much faster using the A2	A2_road_(Great_Britain)	Delay times (positive)

---

[1] http://sentistrength.wlv.ac.uk/

12.8 Experiment Setup and Results

In the following paragraphs we will present the first two stages of the methodology. We will discuss the data-collection approach, the structure of the classifier and the evaluation metrics of the classification process. At the end of the section we will discuss the obtain results.

12.9 Experimental Dataset

Following the ideas presented in Section 12.1 we have assembled an experimental dataset composed of 565,000 tweets from the Portuguese twittosphere. Since we required a training set for our classifier, presented in next paragraph, we manually identified several robots sending traffic-related messages. We considered 3,300 messages posted by these users as positive examples of traffic-related tweets. Then, we randomly picked 41,000 messages from the entire collection to be used as negative examples. We opted for generating an unbalanced data set (ratio of 1 positive example to approximately 12 negative examples) in order to create a pessimist classification model.

12.10 Traffic-Related Tweet Classifier

We implemented a classifier to decide if a text is related to mobility or not. We used Support Vector Machines (SVM) (Cortes and Vapnik, 1995; Vapnik, 1995) as the machine-learning algorithm. SVM is a powerful binary classification algorithm that has proven to be effective in many text classification settings (Joachims, 1998). We used the LibSVM library (Chang and Lin, 2001) available through the Weka (Hall et al., 2009), a software toolkit.

After some preliminary experiments, we decided to configure the SVM algorithm to use a linear kernel function. Simple tests allowed us to check that the performance obtained using polynomial and radial kernels were similar, and thus, did not compensate the extra computational burden. All the remainder parameters were kept in their default values.

Messages were vectored using a unigram bag-of-words approach. Words with less than four characters written in lowercase were considered to be stop words and were removed. We also removed all punctuation characters.

12.11 Evaluation Metrics

The performance of a classifier is usually evaluated using a test set, or using a k-fold cross-validation scheme over the training dataset. However, in our case, the messages we wish to capture are not part of the initial training set, which is composed of traffic-related messages submitted by robot users. Therefore, traditional testing, as k-fold cross-validation, schemes cannot be applied.

Instead, we opted for manually assessing the performance of the classifier: we use the classifier to process a large set of unlabelled data, and we manually verify the results. Messages can be evaluated according to three possible cases: (1) traffic-related and posted by humans; (2) traffic-related but posted by robots; and (3) not related to traffic. We will evaluate the classifier for the task of finding traffic-related messages posted by humans (i.e., case (1)). Traffic related messages posted by robot users would not be considered valid. Thus, true positives (TP) and false positives (FP) are given by:

$$TP_{HMN} = (1) \tag{12.1}$$

$$FP_{HMN} = (2) + (3) \tag{12.2}$$

The number of false negatives FN_{HMN}, that is, the number of valid messages in the collection that were not identified by the classifier, is given as a function of estimate on the total number of traffic messages posted by (human) users, which was found to be 0.05% of all messages.

We can now compute on Precision, Recall, and F-score:

$$P = \frac{TP_{HMN}}{TP_{HMN} + FP_{HMN}} \tag{12.3}$$

$$R = \frac{TP_{HMN}}{TP_{HMN} + FN_{HMN}} \tag{12.4}$$

$$F = 2 * \frac{P * R}{P + R} \tag{12.5}$$

The SVM classifier produces a continuous classification decision value ranging from -1 to 1. Negative decision values correspond to messages that the classifier considered not to be related to traffic, while positive values correspond to messages that are considered being related to traffic. We can impose a minimum threshold on the value for positive decision th_{min} to reduce the number of false positives. We can then compute all the performance measures previously presented for various values of th_{min} and, hence, test the sensitivity of the classifier to this parameter.

12.12 Results Discussion

In this section, we present the results obtained in both stages of the bootstrapping approach. The Tables 12.4 and 12.5 show the absolute and evaluation results achieved by the classification model in the first and second iterations.

In the first iteration (i.e., using the classifier trained only with robot-submitted messages as positive examples), we present the results for several values of the decision threshold, th_{min}, using as test sample of approximately 260,000 unlabelled Twitter messages.

Table 12.4: Absolute results achieved in the second stage of the bootstrapping process.

	1st Iteration		2nd Iteration	
th_{min}	$TP+FP$	TP_{HMN}	$TP+FP$	TP_{HMN}
0.60	62	13	47	21
0.70	50	13	37	20

Tables 12.4 and 12.5 also show the results achieved using the second version of the classification model and another test corpus. The additional labeled data used to build the second version of the classifier was composed by 15 human-generated messages, correctly found in the previous iteration – as positive examples – and 75 messages previously misclassified – as negative examples. The test corpus was composed of approximately the same size as the first one (about 260,000 messages) but with distinct messages.

As it can be seen, the bootstrapping strategy we used led to a significant improvement in the performance of the classifier both in terms of precision and recall. In Table 12.2 we can see that the value of average F-score after bootstrapping was approximately two times higher. In other words, we were able to almost double the average value of precision and recall simultaneously.

Notably, this increase in performance was obtained by adding only a few additional positive (15) and negative (75) examples to the initial training set (3,300 positive examples and 41,000 negative examples). After analysis, we verified that this increase in performance was mostly due to the addition of the negative examples, which allowed building a classification model that was more robust to messages of certain topics. Among these, the most frequent cases were related to Twitter messages about political matters, which tend to share some specific keywords (e.g., "right" and "left"), and also make frequent references to certain locations.

12.13 Conclusions

In this chapter, we discuss a methodology to support not only traffic characterization of a given transportation system, but also decision-making in transportation domain issues as feedback to users experience and opinion. Particular attention we give to the classification

Table 12.5: Precision, Recall, and F-score results in the first and second iteration of the bootstrapping training strategy.

	1st Iteration			2nd Iteration		
th_{min}	$P_1(\%)$	$R_1(\%)$	$F_1(\%)$	$P_2(\%)$	$R_2(\%)$	$F_2(\%)$
0.6	21.0	10.0	13.5	44.7	16.2	23.8
0.7	26.0	10.0	14.4	54.1	15.4	24.0
AVG	23.5	10.0	14.0	49.4	15.8	23.9

stage of the methodology. Here, we presented a two-stage bootstrapping strategy to train a text classifier capable of identifying traffic-related messages posted by users on Twitter. The main challenges at stake were related to the relatively low percentage of relevant messages in the Twitter stream ($< 0.05\%$), which was problematic both for compiling appropriate training sets and for achieving useful classification performance. We trained a first version of the classifier using traffic-related Twitter messages automatically sent by a few official news sources (which were identified manually), and we used this classifier to compile additional positive and negative examples from a large collection of Twitter messages. By expanding the training set with these newly found examples, we were able to train a second version of the classifier, which achieved F-measure of approximately 23% (i.e., an 85% increase in performance in terms of F-measure in relation to the previous classifier). In other words, we show that it is possible to train a relatively accurate text classifier for traffic-related Twitter messages (given the percentage of relevant messages in the universe at stake) with almost no annotation effort. Given the very large number of Twitter messages being posted hourly, we believe that our experiments show that exploring traffic-related information from this media can become an interesting option for complementing information gathered by more traditional sensors (which tend to be limited by several factors such as cost, mobility, availability, etc.).

The data we can gather using discussed methodology can be used to provide a detailed user perspective of several aspects of a mobility network and particular routes. For urban planners this information complements traditional, and objective, metrics. It enables decision-making on the basis of the needs of users in their own perspective. Furthermore, since this information is richer than the traditional metrics, there are additional interventions that can be planned based on it. An urban planner could, for example, decide to fix a particular section of road pavement based on negative sentiment toward the aspect of road conditions. This is something traditional sensors cannot achieve. Since a system such as the one we describe can be in continuous operation, it is also possible to gauge the impact of changes to the network. Taking the previous example of fixing road conditions, one might compare data captured before the intervention, and data captured afterward, thus having a clear indication if the problem was indeed fixed in the eye of the users. The number of mentions to particular segments of the mobility network we wish to study may be an interesting metric in itself, without sentiment. Since we are only dealing with traffic-related posts or tweets, the popularity of these segments may be related to their traffic volumes. A semantic enrichment standpoint of routes, or trajectories, as described by Parent et al. (2012) sentiment would be helpful in explaining route popularity. By enabling further characterization of routes, it would provide an overall better understanding of human mobility for researchers, and allow the development of new applications.

References

Agichtein, E., Castillo, C., Donato, D., Gionis, A., Mishne, G., 2008. Finding high-quality content in social media. International Conference on Web Search and Web Data Mining (WSDM '08). ACM, 183–194, (doi:10.1145/1341531.1341557).

Albuquerque, F.D.C., Barbosa, I., Casanova, M.A., de Carvalho, M.T.M., de Macêdo, J.A.F., 2012. Proactive monitoring of moving objects. Fourteenth International Conference on Enterprise Information Systems 1, 191–194.

Albuquerque, F.D.C., Casanova, M.A., Macedo, J.A.F.D., Carvalho, M.T.M.D., Renso, C., 2013. A proactive application to monitor truck fleets. Fourteenth International Conference on Mobile Data Management (MDM). IEEE 1, 301–304, (doi: 10.1109/MDM. 2013.44).

Asur, S., Huberman, B.A., 2010. Predicting the future with social media. International Conference on Web Intelligence and Intelligent Agent Technology (WI-IAT). IEEE/WIC/ACM 1, 492–499, (doi: 10.1109/WI-IAT.;1; 2010.63).

Bifet, A., Frank, E., 2010. Sentiment knowledge discovery in twitter streaming data. In: Pfahringer, B., Holmes, G., Hoffmann, A. (Eds.), Discovery Science. Springer, Berlin Heidelberg, pp. 1–15.

Boanjak, M., Oliveira, E., Martins, J., Mendes Rodrigues, E., Sarmento, L., 2012. TwitterEcho: a distributed focused crawler to support open research with twitter data. Twenty-first International Conference companion on World Wide Web. ACM, 1233–1240, (doi: 10.1145/2187980.2188266).

Carpenter, T., Way, T., 2012. Tracking sentiment analysis through Twitter. International Conference on Information and Knowledge Engineering.

Carvalho, S., Sarmento, L., Rossetti, R.J.F., 2010. Real-Time Sensing of Traffic Information in Twitter Messages. Madeira Island, Portugal, IEEE ITSC Workshop on Artificial Transportation Systems and Simulation, September 19–22.

Chang, C.C., Lin, C.J. 2011. LIBSVM: a library for support vector machines. ACM Transactions on Intelligent Systems and Technology (TIST), 2(3), 27.

Cheng, Z., Caverlee, J., Lee, K., 2010. You are where you tweet: a content-based approach to geo-locating twitter users. Nineteenth ACM International Conference on Information and Knowledge Management, 759–768, (doi:10.1145/1871437.1871535).

Chesley, P., Vincent, B., Xu, L., Srihari, R.K., 2006. Using verbs and adjectives to automatically classify blog sentiment. Proceedings of AAAI-CAAW-06, the Spring Symposia on Computational Approaches to Analyzing Weblogs. pp. 27–29.

Cortes, C., Vapnik, V. 1995. Support-vector networks. Machine learning, 20(3), 273–297. (doi:10.1023/A: 1022627411411).

Esparza, S.G., OMahony, M.P., Smyth, B., 2012. Mining the real-time web: a novel approach to product recommendation. Knowledge-Based Systems 29, 3–11, (doi: 10.1016/j.knosys.2011.07.007).

Finkel, J.R., Grenager, T., Manning, C., 2005. Incorporating non-local information into information extraction systems by Gibbs sampling. Proceedings of the Forty-third Annual Meeting of the Association for Computational Linguistics (ACL 2005), 363–370.

Hall, M., Frank, E., Holmes, G., Pfahringer, B., Reutemann, P., Witten, I.H., 2009. The weka data mining software: an update. SIGKDD Explorations 11 (no. 1), 10–18, (doi: 10.1145/1656274.1656278).

He, J., Shen, W., Divakaruni, P., Wynter, L., Lawrence, R., 2013. Improving Traffic Prediction with Tweet Semantics. Proceedings of the Twenty-third International Joint Conference on Artificial Intelligence, 1387–1393.

Joachims, T., 1998. Text categorization with support vector machines: learning with many relevant features. Tenth European Conference on Machine Learning, 137–142.

Kaschesky, M., Sobkowicz, P., Bouchard, G., 2011. Opinion mining in social media: modeling, simulating, and visualizing political opinion formation in the web. Proceedings of the Twelfth Annual International Conference on Digital Government Research, 317–326, (doi: 10.1145/2037556.2037607).

Kobayashi, N., Inui, K., Matsumoto, Y., 2007. Extracting aspect-evaluation and aspect-of relations in opinion mining. Proceedings of the 2007 Conference on Empirical Methods on Natural Language Processing and Computational Natural Language Learning (*EMNLP-CoNLL*). pp. 1065–1074.

Kwon, N., Shulman, S.W., Hovy, E., 2006. Multidimensional text analysis for eRulemaking. Proceedings of the international conference on Digital government research. Dig. Gov Soc. N. Am., 157–166, (doi: 10.1145/1146598.1146649).

Lin, W.H., Wilson, T., Wiebe, J., Hauptmann, A., 2006. Which side are you on?: identifying perspectives at the document and sentence levels. Proceedings of the Tenth Conference on Computational Natural Language Learning. Association for Computational Linguistics, 109–116.

Liu, B., 2012. Sentiment analysis and opinion mining. Synthesis Lectures on Human Language Technologies, 5(1), 1–167.

Meij, E., Weerkamp, W., de Rijke, M., 2012. Adding semantics to microblog posts. Fifth ACM International Conference on Web Search and Data Mining. ACM, 563–572, (doi: 10.1145/2124295.2124364).

Parent, C., Spaccapietra, S., Renso, C., Andrienko, G., Andrienko, N., Bogorny, V., Damiani, M.L., Gkoulalas-divanis, A., Macedo, J., Pelekis, N., Theodoridis, Y., Yan, Z., 2012. Semantic trajectories modeling and analysis. ACM Comp. Surv., (doi: 10.1145/2501654.2501656).

Rao, D., McNamee, P., Dredze, M., 2013. Entity linking: Finding extracted entities in a knowledge base. Multi-source, Multilingual Information Extraction and Summarization. Springer Berlin Heidelberg, pp. 93–115.(doi: 10.1007/978-3-642-28569-1_5).

Ritter, A., Clark, S., Etzioni, O., 2011. Named entity recognition in tweets: an experimental study. In Proceedings of the Conference on Empirical Methods in Natural Language Processing. Association for Computational Linguistics, 1524–1534.

Sakaki, T., Matsuo, Y., Yanagihara, T., Chandrasiri, N.P., Nawa, K., 2012. Real-time event extraction for driving information from social sensors. International Conference on Cyber Technology in Automation, Control, and Intelligent Systems. IEEE, 221–226, (doi: 10.1109/CYBER. 2012.6392557).

Sakaki, T., Okazaki, M., Matsuo, Y., 2010. Earthquake shakes twitter users: real-time event detection by social sensors. New York, NY, USA. ACM, Nineteenth International Conference on World Wide Web, pp. 851–860 (doi: 10.1145/1772690.1772777).

Sasaki, K., Nagano, S., Ueno, K., Cho, K., 2012. Feasibility Study on Detection of Transportation Information Exploiting Twitter as a Sensor. Sixth International AAAI Conference on Weblogs and Social Media.

Schweitzer, L., 2012. How Are We Doing? Opinion Mining Customer Sentiment in US Transit Agencies and Airlines via Twitter. Transportation Research Board Annual Meeting.

Shulman, S., Hovy, E., Callan, J., Zavestoski, S., 2005. Language processing technologies for electronic rulemaking: A project highlight. National conference on Digital government research. Dig. Gov. Soc. N. Am., 87–88.

Stylios, G., Christodoulakis, D., Besharat, J., Vonitsanou, M.A., Kotrotsos, I., Koumpouri, A., Stamou, S., 2010. Public opinion mining for governmental decisions. EJEG 8 (2), 203–214.

Thelwall, M., Buckley, K., Paltoglou, G., Cai, D., Kappas, A., 2010. Sentiment strength detection in short informal text. J. Assoc. Inf. Sci. Technol. 61 (Issue 12), 2544–2558, (doi: 10.1002/asi.v61:12).

Tsytsarau, M., Palpanas, T., 2012. Survey on mining subjective data on the web. J. Data Min. Knowl. Disc. 24 (3), 478–514, doi: 10.1007/s10618-011-0238-6. (doi: 10.1007/s10618-011-0238-6).

Tumasjan, A., Sprenger, T.O., Sandner, P.G., Welpe, I.M., 2010. Predicting elections with twitter: What 140 characters reveal about political sentiment. Fourth International AAAI Conference on Weblogs and Social Media. AAAI, 178–185.

Vapnik, V.N., 1995. The nature of statistical learning theory. Springer-Verlag, New York, Inc.

Index

Printed in the United States
By Bookmasters

Printed in the United States
By Bookmasters